Dr Riemann's Zeros

Karl Sabbagh

Atlantic Books
London

Frontispiece photographs:

ROW 1: *Tom Apostol, Michael Berry*
ROW 2: *Enrico Bombieri, Louis de Branges, Daniel Bump, Alain Connes*
ROW 3: *Brian Conrey, Martin Huxley, Alexander Ivic, Henryk Iwaniec, Matti Jutila*
ROW 4: *Yoichi Motohashi, Nikolai Nikolski, Andrew Odlyzko, Charles Ryavec*
ROW 5: *Peter Sarnak, Atle Selberg*

First published in Great Britain in 2002 by Atlantic Books,
an imprint of Grove Atlantic Ltd

9 8 7 6 5 4 3 2 1

A CIP catalogue record for this book is available from the British Library.

ISBN 1 84354 100 9

Printed in Great Britain by CPD, Ebbw Vale, Wales

Atlantic Books
An imprint of Grove Atlantic Ltd
Ormond House
26–27 Boswell Street
London WC1N 3JZ

[The Riemann Hypothesis] is now unquestionably the most celebrated problem in mathematics and it continues to attract the attention of the best mathematicians, not only because it has gone unsolved for so long but also because it appears tantalizingly vulnerable and because its solution would probably bring to light new techniques of far reaching importance.

H.M. Edwards, mathematician and author of *Riemann's Zeta Function*

CONTENTS

ACKNOWLEDGEMENTS

This book started as a result of a number of very enjoyable conversations with Bela Bollobas about what sort of book it would be possible to write to give non-mathematicians a glimpse of the pleasures and importance of mathematics. As a result I decided to concentrate on mathematical problems, and one problem in particular – the Riemann Hypothesis.

The core of the book has emerged from conversations with two dozen or so mathematicians who are currently working on the Riemann Hypothesis and who gave me their time, sometimes on more than one occasion, to talk not only about the hypothesis itself but about their own passion for mathematics, the effect mathematics has had on their lives, how they see the philosophical significance of pure mathematics and whatever else occurred to me to ask them at the time.

Because of his single-minded focus on the problem and his firm belief in the possibility of a solution, I went deeper into mathematics with Louis de Branges than with the other mathematicians I spoke to, and I thank him for the time and hospitality he supplied so generously on several occasions.

Most of the other mathematicians I consulted are quoted in the book, but even those on the following list who are not quoted also contributed immeasurably to my understanding of the subject.

They are: Tom Apostol, Michael Berry, Enrico Bombieri, Daniel Bump, Alain Connes, Brian Conrey, Louis de Branges, David Farmer, Steve Gonek, Andrew Granville, Roger Heath-Brown, Martin Huxley, Alexander Ivic, Henryk Iwaniec, Matti Jutila, Jonathan Keating, Xian Jin Li, Hugh Montgomery, Yoichi Motohashi, Nikolai Nikolski, Andrew Odlyzko, Samuel Patterson, Charles Ryavec, Peter Sarnak and Atle Selberg.

In addition to talking to me about his work, Michael Berry also took time to read the manuscript and correct some of my more egregious mathematical errors. Peter Wilson read the manuscript to give me a layman's reactions to the story. I thank them both.

I would also like to thank Adrian Moore, who, although not a mathematician but a philosopher, helped me to understand some of the philosophical issues underlying the topic.

Stephen Bann drew my attention to the passage on imaginary numbers from *Young Törless* by Robert Müsil, and Artie Shaw gave me permission to quote from his memoir *The Trouble with Cinderella*.

I would like to thank the American Institute of Mathematics and its director Brian Conrey for allowing me to sit in on their workshop on L-functions at Palo Alto. I also appreciated very much the opportunity to attend part of a workshop on the theory of Riemann zeta functions at the unique Mathematisches Forschungsinstitut in Oberwolfach in Baden-Württemburg.

As ever, the London Library was a rich source of information. Its mathematical collection is entirely arbitrary (it is largely a humanities library), but the books that it does have are a mine of unusual and sometimes eccentric mathematical material.

I have also found invaluable the MacTutor History of Mathematics Archive organized by the School of Mathematics and Statistics of the University of St Andrews in Scotland, to be found on the Internet at http://turnbull.dcs.st-and.ac.uk/~history/index.html.

At Farrar Straus Giroux in New York, John Glusman, a man with no previous appetite for mathematics, nevertheless agreed enthusiastically to commission this book, and was followed equally enthusiastically in the UK by Toby Mundy at Atlantic Books. Both of them have provided the unstinting support that every author needs and doesn't always get. Clara Farmer and Rebecca Wilson (at Atlantic) and John Woodruff have all shared the task of making the book as comprehensible as possible to a non-mathematician, through their advice on text, diagrams and mathematical expressions, and I thank them all.

Finally, I would like to remember the two men who taught me mathematics at school, Bill Blight and Joe Church. For a year or two, with their help, I thought I understood what it was all about, and glimpsed the fascination that, at a much higher level, drives professional mathematicians the world over.

THE MATHEMATICIANS

TOM APOSTOL, California Institute of Technology

MICHAEL BERRY, University of Bristol

ENRICO BOMBIERI, Institute for Advanced Study, Princeton

DANIEL BUMP, Stanford University, California

ALAIN CONNES, Institut des Hautes Études Scientifiques, Paris

BRIAN CONREY, American Institute of Mathematics, Palo Alto, California

LOUIS DE BRANGES, Purdue University, Lafayette, Indiana

STEVE GONEK, University of Rochester, New York

ANDREW GRANVILLE, University of Georgia, Athens, Georgia

ROGER HEATH-BROWN, University of Oxford

MARTIN HUXLEY, University of Cardiff

ALEXANDER IVIC, University of Belgrade, Serbia

HENRYK IWANIEC, Rutgers University

MATTI JUTILA, Turku University

JONATHAN KEATING, Bristol University

XIAN JIN LI, Brigham Young University, Utah

HUGH MONTGOMERY, University of Michigan

YOICHI MOTOHASHI, Nihon University, Tokyo

NIKOLAI NIKOLSKI, University of Bordeaux

ANDREW ODLYZKO, University of Michigan

SAMUEL PATTERSON, Mathematisches Institut, Universitäts Göttingen

CHARLES RYAVEC, State University of California at Santa Barbara

PETER SARNAK, Institute for Advanced Study, Princeton

ATLE SELBERG, Institute for Advanced Study, Princeton

A NOTE ABOUT THE TOOLKITS

At the end of the book I have included what I call 'toolkits' for readers who would like to be introduced to – or refresh their memories about – some basic mathematical terminology, symbols and methods. When I mention logarithms for the first time, for example, if you already know what a logarithm is you can just continue reading. But if you want to understand them a little more you can turn to the appropriate toolkit. Similarly, I have tried to cover mathematical ideas such as 'equations', 'graphs', 'infinite series' and so on in just enough detail to make the main story comprehensible but without trying to write a mathematical textbook.

PROLOGUE

Many people would say that the task I am embarking on by writing this book is doomed to failure. For them, it is as if I have chosen to write a book in Arabic in the belief that non-Arabic speakers will find something to enlighten them within its pages. I prefer a different analogy. For me, it will be as if I am describing a remote tribe whose customs and language are unfamiliar to the reader, but whom I understand enough to convey something of their inner and outer lives.

To read anthropologists such as Bronoslaw Malinowski on the Magic Gardens of the Trobrianders or Edward Evans Pritchard on the oracles of the Nuer may, for some people, be so far from their experience or interests that such books might as well be in Arabic. But for others at least, the insights that anthropologists can have into the minds of their subjects are possible to appreciate, whether or not the reader speaks or reads the languages of those tribespeople.

Mathematicians also form a kind of tribe, with its own language and customs. They have a veneer of civilization, the type of 'civilization' that forms the framework of the lives of the rest of us. They go to work in offices or departments, do the washing up, walk the hills, and – not very often – watch television. But their inner lives are very different from those of the rest of us. It is given to them to see *truths* with a clarity that is sometimes breathtaking. As we will discover, there is some argument over whether the stuff of mathematics is *out there* – facts and relationships waiting to be discovered – or *in here* – creations of the human mind that are akin to inventions, paintings or poems. If you believe – as I do – that mathematical truths are out there, then a mathematician proving a hypothesis is discovering something which is true for ever and a truth that he or she may well be the first to appreciate. Living on the surface of a planet which is rapidly running out of unexplored territory, *mathematical* explorers believe that there is no limit to the new discoveries they might make. It seems, if the history of mathematics is anything to go by, that they will never run out of territory to explore, new discoveries to savour. But just as modern social anthropology teaches us how much we have in common with supposedly exotic peoples, I hope this book will suggest that mathematicians are not as different from the rest of us as we might think.

I was recently in a bookshop in a small American university town when I came across two shelves of maths books. These were not basic textbooks but highly specialized works with titles like *Finite Markov Chains* and *Rings, Fields and Vector Spaces*. For me, maths books – even those written at a level far above my comprehension – have an irresistible appeal, partly because of their titles, partly because I remember enough of my own university maths to understand small sections of text that appear as patches of blue in a generally cloudy sky. I bought one book largely for its title, as incorrect grammatically as its content was – I'm sure – correct mathematically. It was called *Linear Algebra Done Right.*

And then I saw it, a book that proved that mathematicians are human – a book called *The Joy of Sets*, in homage to Alec Comfort's book, *The Joy of Sex*. Now, if I'd seen a book by a marine archaeologist called *The Joy of Wrecks* or one by an explorer called *The Joy of Treks*, I wouldn't have given it a second glance. Archaeologists and explorers are members of the human race, like you and me. We could imagine ourselves doing their jobs, sharing a joke with them, hearing about their day under water or on a steep hillside, and talking about sex.

But mathematicians? Come on. Then I realized that that is what my book has to be about – the humanity of mathematicians. Not humanity in the sense that they love all mankind, give to charity and help around the house, but humanity in the sense of human-ness. The mathematical brain may be different from the non-mathematical brain, but not very – certainly no more different than a footballer's brain is from mine, or mine from any woman's. The mathematician immersed in a problem is not on another planet, merely as preoccupied as other people can be with a piece of needlework, a wood carving or a béchamel sauce. But his or her life outside the mathematical enterprise, the passions, disappointments and frailties, have much in common with yours or mine.

Many people are so afraid of mathematics, panicking when faced with mathematical expressions or statements, that they ascribe almost superhuman powers to those who *are* at home with such ideas. Part of this reaction is based on an inability to believe that this stuff can be at all interesting. But mathematicians have a secret. Not only is mathematics interesting, the insights of mathematics are arguably the *most* interesting discoveries ever made about the universe.

But this is not a jolly, 'maths is really fun' book for reluctant undergraduates. I would be incapable of writing such a book since I don't know enough about the subject. Instead, it's a portrait of a particular time in mathematics, when one of maths' most important unsolved problems might be on the brink of being solved. It's a problem that has been around for nearly a hundred and fifty years, and every pure mathematician worth his or her salt has longed for it to be settled one way or another. In fact, it may actually be insoluble, and even knowing *that* would be an important step in mathematics. The task is to prove the truth of a particular mathematical statement, known as the Riemann Hypothesis, after Bernhard Riemann, the mathematician who first stated it in a paper published in 1859. It's a statement which, if true, will reveal a secret about prime numbers – those numbers such as 13 and 41 that cannot be divided by any other number (except, of course, 1) without leaving a remainder.

Proving the Riemann Hypothesis would not be like splitting the atom or cracking the genetic code. A solution has no foreseeable practical consequences that could change the world. But then neither do the novels, paintings and philosophical writings we might admire from some of the world's great creators. As we'll see, there *could* be practical applications of the type of mathematics that goes into investigating the Riemann Hypothesis, but that's not why mathematicians do it. Many of them get more pleasure out of mathematics than almost any other activity. And they often first discovered this pleasure when they were young.

Most mathematicians did not just take maths up 'as a job', in the way that someone becomes a banker or a publisher or a lawyer. For few of *these* professions did the serious intellectual bases of their discipline begin at an early age. Not many children say 'I want to be a banker (or a publisher or a lawyer) when I grow up'. Yet for the mathematicians I met, almost all of them did their first serious thinking about the subject when they were in their early teens, or even younger.

'I was always interested in math even as a young kid,' said Brian Conrey. 'I remember my father telling me about negative numbers when

I was four or five and just being fascinated with this. Then, maybe when I was in about eighth or ninth grade, I actually had gotten interested in number theory and I'd checked out a book from my school library – *Number Theory and Its History*. And I read through and saw this problem about twin primes,* "Are there infinitely many twin primes?" and I couldn't believe that we didn't know the answer to that question – I said, "That can't be that hard", and I started working on it. I actually thought that I had solved it as a high-school student, but I took it to some professors at the University of New Mexico, and they were interested in it and pointed out where the mistake was.'

Charles Ryavec had a similar childhood experience. 'I remember in the tenth grade, we had this math teacher who separated everybody according to rules – there were the A students, the B students, and so forth, and in school I was more or less a C or D math student, which put me by the window. But all his neat books were by the window and I could reach over to them. And I found this book by Klein on the icosahedron and the solution of equations of the fifth degree. And I remember I'd spend all day trying to solve a cubic equation and I never was able to do it. And so I went to the library and I found a book where they actually solved a cubic, and I think that was probably the first big thing where I really loved maths, seeing that.'

Alexander Ivic, a Serbian mathematician from Belgrade, thought he had solved an important mathematical problem when he was in his teens. In this case, he actually had, but so had someone else before him. 'I discovered the Lagrange interpolation formula for polynomials when I was fifteen or sixteen years old,' he said, 'and then for about a week I thought, "This is it, this is the greatest discovery of all time." When I found out in a book that it was known for two hundred and fifty years at least, then I wisely decided to give up research until I was ready for it. But already at that age I knew I was going to be a professional mathematician.'

For people who never developed a talent for mathematics, these statements will arouse no spark of empathy or recognition. It's not just the display of facility with numbers, it's the fact that these people clearly enjoyed their first contact with mathematics. Louis de Branges, who is now a mathematician working on the Riemann Hypothesis, first

---

* Prime numbers separated by 2, e.g. 11 and 13 or 29 and 31.

became fascinated by mathematics when a friend of the family, Mr Irénée Du Pont, the former president of the Du Pont Company, posed a problem for him when he was a child.

De Branges related how he caddied for his grandfather who played golf on Sunday mornings with du Pont. 'Mr du Pont always drank a glass of rum and orange juice in the clubhouse after playing golf. One morning he showed an unexpected interest in my mathematical education by posing a problem: Find positive integers a, b and c such that

$$a^3 + b^3 = 22c^3$$

I spent the fourth form year solving it.'[1]

Many of these young mathematicians were undaunted by the challenge of competing with older and wiser minds, even to the extent of disagreeing with the textbooks. Julia Robinson's starting point was the square root of 2. Some square roots of whole numbers are themselves whole numbers: $\sqrt{25}$ is 5, for example. But some are decimal numbers: $\sqrt{2}$ is 1.414 213 562..., with the dots meaning, as they will throughout this book, that you can extend a decimal fraction (or a series) for ever. In this case it means that the more digits you take before squaring the number, the nearer the result will be to 2. But only by multiplying an infinite number of decimal digits together could you produce exactly 2. Robinson wrote:

> I first became interested in mathematics when our textbook in the seventh or eighth grade claimed that no matter how far the decimal expansion of $\sqrt{2}$ was carried out it would never become periodic.* I didn't see how anyone could know that – all they could know was that the expansion had not become periodic in the part that was calculated. I decided to expand it a long ways in hopes of finding a period when I got home. Well, the arithmetic defeated me after a bit even though I, like most mathematicians, am quite stubborn.[2]

One of the strangest stories about a childhood interest in mathematics was told to me by Alain Connes, a brilliant French mathematician who devised his own system of mathematics for solving the problems he was presented with in class. 'What happened was that I first began to work

---

* i.e. start repeating itself, e.g. 1.414 213 562 562 562 562 562 562...

in a very small place in the mathematical geography. This meant that I had a strange and polarized view of things which was peculiar to me, and somehow from then on I travelled through many other places, but without ever breaking this thread which I had created from the beginning and which was giving me a particular point of view, quite different from the standard point of view. So I had my own system, which was very strange because when the problems that the teacher was asking fell into my system then of course I would have an instant answer, but when they didn't – and many problems, of course, didn't fall into my system – then I would be like an idiot, and I wouldn't care.'

It's easy to dismiss these examples of early mathematical interest as rare and exotic examples of child prodigies. But there are too many of them to explain them in this way. Almost every mathematician I have spoken to has a story like these – examples of children with no particular training or expertise, no parental or school pressure, and no obvious difference from their playmates and school friends, understanding a mathematical statement well enough to glimpse its depth. What's more, the children have experienced the pleasure and excitement that can come from pursuing the implications of the statement to its logical conclusion.

For a select few of them, such as the ones I follow in this book, their early curiosity has matured into a passion to prove the most difficult problem any of them has ever tackled, a passion which is largely pursued for its own satisfaction. I say 'largely' because although people have pursued a proof of the Riemann Hypothesis for a hundred years or more purely for the intellectual excitement it offers, there is now a prize of $1 million donated by the Clay Institute for the first person to prove it.

In its publicity directed at mathematicians who might wish to try their hand at proving the Riemann Hypothesis, the Institute describes the problem with a disarming mixture of primary school arithmetic and specialized mathematical terminology:

Some numbers have the special property that they cannot be expressed as the product of two smaller numbers, e.g. 2, 3, 5, 7, etc. Such numbers are called prime numbers, and they play an important role, both in pure mathematics and its applications. The distribution of such prime numbers among all natural numbers does not follow any regular pattern. However, the German mathematician G. F. B. Riemann (1826–1866) observed that the frequency of prime numbers is very closely related to the behavior of an elaborate function '$\zeta(s)$'* called the Riemann Zeta function. The Riemann Hypothesis asserts that all interesting solutions of the equation

$$\zeta(s) = 0$$

lie on a straight line. This has been checked for the first 1,500,000,000 solutions. A proof that it is true for every interesting solution would shed light on many of the mysteries surrounding the distribution of prime numbers.[3]

'Functions' and 'equations' are concepts that will be unfamiliar to most people, but they are meat and drink to mathematicians, whose attempts to convey these abstract concepts in concrete terms, through metaphors and analogies, can help the non-mathematician to glimpse the depth of the concepts that were behind Bernhard Riemann's big idea.

Why is it so important? It's enough at this stage to say that a proof of the Riemann Hypothesis would, at a stroke, tell mathematicians a huge amount about an important class of numbers – the prime numbers, which dominate the field of pure mathematics. What's more, over the years, attempts to prove the Riemann Hypothesis have taken mathematicians into all sorts of unexpected areas of mathematics, sometimes leading them to create new mathematical tools with which to attack the problem. The story spans several hundred years and provides an illustration of something that happens time and again in mathematics: a question is asked which has a limpid simplicity about it, but that simplicity evaporates very quickly as the question assumes a fearsome and unexpected complexity. Centuries before Riemann considered the problem, mathematicians had asked how the prime numbers are distributed throughout the whole numbers. The early attempts to answer this question led quickly to the realization that there was something special about the prime numbers, something that resisted the efforts of mathematicians to answer even quite simple questions about them.

It is this aspect of the prime numbers, culminating in the most intractable problem of all, that so fascinates mathematicians.

I once asked Michael Berry, one of the people whose work is making

---

* Pronounced 'zeta of s'.

a significant contribution to understanding the Riemann Hypothesis, what it felt like to be sixty.

He replied, 'Sixty, sandwiched between two primes, has the property that no smaller number has more distinct prime factors. Other than that, nothing special.'

An everyday familiarity with prime numbers is second nature to many mathematicians working in number theory, the field of mathematics that deals with prime numbers and the other whole numbers. And to understand the importance of the Riemann Hypothesis in modern mathematics there is only one place to start – with these building blocks of the entire system of whole numbers. It's as obvious as 1, 2, 3, ...

1 Prime time

Mathematicians have tried in vain to this day to discover some order in the sequence of prime numbers, and we have reason to believe that it is a mystery into which the human mind will never penetrate.

Leonhard Euler

The Riemann Hypothesis emerged from the attempts of mathematicians to understand the subtleties of prime numbers. These are whole numbers, also called integers, that cannot be divided by smaller integers without leaving a remainder. They are fundamental to our number system, as Jon Keating, a mathematician working at the University of Bristol, explains:

'Primes are like pieces of Lego. You have individual blocks of Lego which you can't break down any further. The smallest blocks of Lego come in different sizes but you can't break them in half. They're the primes. Out of those blocks you can build buildings, you can build Lego objects. Those are like all the other numbers, the non-primes. One question is, is a brick itself a Lego object? I would say yes, and therefore in that sense the primes are different from the rest of the numbers, but they are part of the numbers – they are the numbers you can build everything out of and they're the numbers you can't break down any further.'

Prime numbers are the ones printed in bold in the following list. It's a list with which everyone is familiar – the counting numbers:

1, **2**, **3**, 4, **5**, 6, **7**, 8, 9, 10, **11**, 12, **13**, 14, 15, 16, **17**, 18, **19**, 20, 21, 22, **23**, 24, 25, 26, 27, 28, **29**, 30, **31**, 32, 33, 34, 35, 36, **37**, 38, 39, 40, **41**, 42, **43**, 44, 45, 46, **47**, 48, 49, 50, 51, 52, **53**, 54, 55, 56, 57, 58, **59**, 60, **61**, …

The significance of the list is less in the primes than in the other numbers. All the others can be obtained by multiplying together some combination of prime numbers from earlier in the list. For example, 6 is

2×3. But you can't get any of the prime numbers by multiplying other numbers together. You can't 'split' 19 into the product of smaller numbers ('product' in mathematics meaning the result of a particular multiplication).

There is a chemical analogy that might help to underline the importance of prime numbers. It involves the difference between atoms and molecules. Atoms are the basic building blocks of all matter in the universe. There are 92 naturally occurring elements, from hydrogen to plutonium, each with a characteristic type of atom. These atoms have the ability to stick together, either in collections of atoms of the same element – say, lots of carbon atoms joined together – or of different elements – hydrogen and oxygen, say. A collection of atoms – whether the same or different – linked firmly together is called a molecule. There are many different types of molecule, far more than 92. Although not infinite in number, molecules come in so many different shapes and sizes – because 92 types of atom can combine in many different ways – that for practical purposes they are uncountable.

Any substance, from wood to water, cheese to chalk, can be described by a formula which indicates the exact atomic make-up of each molecule of the specific substance. To get a molecule of water, you take two atoms of hydrogen and attach them to one atom of oxygen, and so the molecule is described as H_2O. Larger molecules are more complex. The mineral called marialite has the formula $Na_4Al_3Si_9O_{24}Cl$, which means that a molecule of marialite is made up of four atoms of sodium (Na) joined to three of aluminium, nine of silicon, twenty-four of oxygen and one of chlorine.

Now, the whole numbers, sometimes called the counting numbers, that we all use every day also fall into two types, a little like atoms and molecules. The trouble is, we don't usually realize this because the two types are mixed up. It's as if water, salt and haemoglobin (molecules) were included in the same list as hydrogen, carbon and iron (atoms). The two types of whole number are the prime numbers and the non-primes, also called composite numbers, just as hydrogen could be seen as 'prime' and water as 'composite'.

In both cases, each composite member is a unique arrangement of the units that make it up. Just as a specific molecule, say of sulphuric acid, is made up of a unique set of atoms – H_2SO_4, or two hydrogens, one sulphur and four oxygens – so a number that is not prime, such as

108,045, is made up of a unique grouping of prime numbers – in this case two 3s, one 5 and four 7s multiplied together, which you could call $3^2 \times 5 \times 7^4$, almost like a formula for this particular composite number. There is no other combination of atoms that will make sulphuric acid, and there is no other combination of primes that will make 108,045. This fact about the whole numbers, which can be proved to be true, is known as the Fundamental Theorem of Arithmetic. But there is one important difference: the atoms run out before you get to a hundred, whereas the primes go on for ever. However large a prime number you discover, I can always show you a larger one.

The infinity of the primes was one of the earliest facts to be established about these special numbers. Since the first known proof, by Euclid in the third century BC, mathematicians have come up with hundreds of important statements about subtler aspects of these numbers, based on analysing how they are actually spaced out along what's called the number line, which stretches from its starting point at zero towards infinity.

Imagine the numbers that we are all familiar with, laid out in a straight line from zero to infinity. You could visualize it as a long straight road with a neat line of houses continuing into the distance, each house numbered in order from 1 to infinity. Now, some houses have one person living in them, and others more than one. The one-person houses are 'primes', and their addresses are prime numbers. The multi-occupant houses have people living in them who are all related to someone living in a prime house. There's a simple rule that determines which family members are living where. If the address is divisible by the address of a 'prime' house, then there's someone from that family living there. For example, house number 6 has two people in it: one from the same family as the resident of number 2 and one related to the inhabitant of number 3.

Number 3 has an address that is not divisible by any other address, so it just has one person, a member of the 3 family living there. And number 4 has two members of 2 living in it, because 4 is divisible by 2 twice, since 4 is 2^2. But number 12 has people from the 2 family and the 3 family living there – two members of 2 and one of 3.

It may seem a cumbersome analogy, but it does have its uses. Along the number street every house has at least one inhabitant, and as you go along the street the houses have to accommodate more and more people because their addresses are divisible by more and more primes. But it turns out that as you walk along the road, passing houses with more and more people living in them, there are still occasional one-person houses scattered here and there. However far you travel, there seem to be members of new families ahead, new primes living in solitary comfort.

As mathematicians have explored this strange road they've found a lot of surprising things. For example, there doesn't seem to be a regular pattern to the distribution of the primes, though there are certain regularities. There is never a situation – apart from houses 2 and 3 – where two primes live next door to each other. But there are many instances – in fact, there are believed to be an infinite number – of two primes separated by one multi-occupant house. This, known as the Twin Primes conjecture, is just one of many hypotheses in number theory which are simple to state but have resisted all attempts at a proof.

There are some long stretches of multi-occupancy houses with no primes at all. In fact, you would find that however long a stretch you care to suggest (a hundred houses, a thousand houses,…), somewhere along this road will be a stretch of multi-occupancy houses as long as that, uninterrupted by primes. There doesn't seem to be a rule about how many people live in adjacent houses. The sole resident of number 10,007 (a prime) lives next door to number 10,006, containing one person from family 2 and one from family 5,003; and next to them is a house, number 10,005, that is home to four residents, one each from families 3, 5, 23 and 29.

Looking at the straight line of whole numbers, the primes get farther apart as you move along the line towards infinity. This suggests that you might eventually reach a number that is the last prime number – from then on, perhaps, every number can be obtained by multiplying together some combination of lower numbers. But, in fact, that doesn't happen. However far you go along the road there will always be more primes ahead, even if they are separated more and more by multi-occupancy houses. Since Euclid's proof of this fact, there have been many different proofs of the infinity of the primes. Let's look at one of them.

Suppose we assume that there *is* a largest prime number – call it L. If we use that assumption as a starting point, and then, by a series of logically impeccable steps, find a way to create a prime number larger than L, this would show that our initial assumption must have been wrong. If I say 'all roads lead to Rome', set off down any old road, and arrive in Florence, I have *proved* that 'all roads lead to Rome' is a false statement. So if we can find a way of making a prime number larger than *L*, we will have shown that there is no number *L* that is the largest prime number. In doing this we will have carried out a particular type of proof called a *reductio ad absurdum*. Here's one way of doing it.

We start with our assumed largest prime, *L*, and multiply it by all the prime numbers smaller than *L* to get a number – call it *C* – that is clearly not prime but composite, since it can be divided by any of the smaller prime numbers. We now add 1 to this composite number. Now, is this new number, *C*+1, prime or composite? For it to be prime there must be no numbers that divide it exactly. The way to discover whether this is so is to divide it, one by one, by all smaller prime numbers. (We need only divide by primes because it would be impossible to find a composite number that could divide exactly into *C*+1 without one of the prime factors of that composite also dividing it. If 21 divides into 231, for example, then so do 3 and 7.) Now, every time we divide *C*+1 by a smaller prime we get an answer that is not a whole number. To make this clear, suppose that someone had said that 17 is the largest prime. If we multiply all the smaller primes together we get $2 \times 3 \times 5 \times 7 \times 11 \times 13 = 30,030$. Add 1, and we get 30,031. However we look at it, 30,031 is a prime because no lower primes – and therefore no lower numbers – will divide into it exactly. Try dividing by 7, for example:

$$\frac{2 \times 3 \times 5 \times 7 \times 11 \times 13 + 1}{7} = 2 \times 3 \times 5 \times 11 \times 13 + \frac{1}{7}$$

In a similar way, dividing by any other smaller prime number will leave a remainder – of $\frac{1}{2}, \frac{1}{3}, \frac{1}{5}, \frac{1}{11}$ or $\frac{1}{13}$. So *C*+1 must be a prime, and it must be one that is larger than *L*.

To recap, having assumed there was a largest prime number, *L*, we produced a contradiction – a prime number that was larger still, *C*+1. And the method we've used would apply to any largest prime number you care to suggest. So this means that our original assumption was wrong: there is no largest prime number – they just go on and on for ever.

Of course, this doesn't mean there isn't a largest *known* prime number. There will always be one of those, for as long as people with time on their hands choose to work out a larger one or, these days, choose to program ever more powerful computers to do the search. In 1876, the new record-holder became

$$170,141,183,460,469,231,731,687,303,715,884,105,727$$

It has 39 digits and remained the largest prime until, in 1951, thanks to a roomful of valves and circuits that comprised one of the earliest digital computers, a larger prime number was discovered. This one had 44 digits, but it didn't hold the record for long, as prime numbers with 79, 157, and then several hundred digits were discovered.

According to a news cutting[4] (shown in Figure 1) which turned up among the papers of the American mathematician Julia Robinson, a larger prime was discovered by a Chicago mathematician. There's something odd about this cutting. The number Dr Krieger found is larger than the 44-digit discovery made using the primitive computer, and yet it appears he did this merely with pencils and paper. I haven't checked the primeness of this number (I leave that as an exercise for the reader, as they say in maths textbooks), but the only other fact I can establish about Dr Krieger is that in 1938 he reported

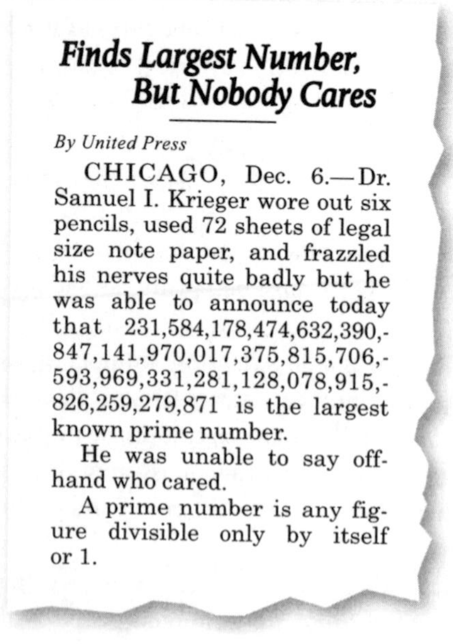

Finds Largest Number, But Nobody Cares

By United Press

CHICAGO, Dec. 6.— Dr. Samuel I. Krieger wore out six pencils, used 72 sheets of legal size note paper, and frazzled his nerves quite badly but he was able to announce today that 231,584,178,474,632,390,-847,141,970,017,375,815,706,-593,969,331,281,128,078,915,-826,259,279,871 is the largest known prime number.

He was unable to say off-hand who cared.

A prime number is any figure divisible only by itself or 1.

FIGURE 1

Mathematics hits the headlines – one of a regular series of news stories about the latest largest prime number.

a counter-example to Fermat's Last Theorem, an example that proved the theorem wrong. Over fifty years later, the English mathematician Andrew Wiles proved the theorem correct.

Since the early days of computers, finding large primes has become little more than a harmless pastime for anyone with a PC. The search has lost its glamour, and larger ones are emerging all the time. In 1998 Roland Clarkson, a nineteen-year-old California student, discovered a new prime, $2^{3,021,377} - 1$, which has 909,526 digits. A mass computer project called GIMPS (Great Internet Mersenne Prime Search)* makes use of the down time on thousands and thousands of PCs around the world, and the largest known prime number, as I write, is $2^{13,466,917} - 1$. It has more than 4 million digits, and if you print it out in type this size, it would be 5 kilometres long.

For some mathematicians, even those passionately interested in primes, the very idea of devoting time to finding the largest prime at any one time is not worth the effort.

'Mathematically this has almost no interest,' said Andrew Granville of the University of Georgia. 'I would be recognized as one of the world experts on such techniques and I'm very interested in how you would go about finding a very large prime number, but it's almost irrelevant when you're doing it except that it can be done. But what happens is every time somebody finds a new largest prime number, it's a cheap market for them. They figure people like to hear this stuff. For people who like a little maths in their lives it really cheapens the subject because it's not very relevant, the actual calculations. There are some calculations that are fantastic and well worth knowing about, but finding the largest prime number…?'

There's a retired engineer called Harvey Dubner who would disagree profoundly with Granville's view. He spends his days finding ever more curious and unusual primes just for the sake of it, like a lepidopterist looking for more and more exotic butterflies. To write these primes, a new notation is needed to save ink and paper. (Mathematicians are always happy to invent new notations if it simplifies things.) If several digits are repeated, either singly or in groups, they are put in

---

* www.mersenne.org/prime.htm A Mersenne prime is a prime number of the form $2^n - 1$, named after the 17th-century French monk, Marin Mersenne, who encouraged mathematicians to investigate these special primes, first studied by Euclid.

brackets with a subscript number showing how often they are repeated. So, for example, the number 144,444,444,444,444,444,443 could be written more succinctly as $1(4)_{19}3$, and 723,232,323,231 could be written as $7(23)_51$. One of Dubner's goals is to find primes with repeated digits. Here are a couple of his recent discoveries.[5] The largest known prime with all digits equal to 0 or 1 is $(1)_{2,700}(0)_{3,155}1$. Just to make clear the magnitude (and perhaps the futility) of this discovery, this number starts with 2,700 1s followed by 3,155 0s, and finishes with another 1.

The prime with the largest number of 0s in it known so far is $134,088 \times 10^{15,036} + 1$. In our new notation this can be written as $134,088(0)_{15,036}1$. This number would take up about eight pages of a normal printed book, and all those pages would be covered with 0s, apart from five digits at the beginning of the first page and a 1 at the end. And remember, in addition to looking odd when written out, this is a prime, so it can be divided by no other number.

But there are other much more interesting questions posed by the existence and distribution of prime numbers, many of which are extremely simple to state and obvious to demonstrate, but seem impossible to prove, such as:

Is there always at least one prime number between successive squares?

It's very easy to test this, and it's certainly true for the first few squares, as this list shows (with primes in bold and squares in regular type):

1, **2**, **3**, 4, **5**, **7**, 9, **11**, **13**, 16, **17**, **19**, **23**, 25, **29**, **31**, 36, **37**, **41**, **43**, **47**, 49, **53**, **59**, **61**, 64, **67**, **71**, **73**, **79**, 81, **83**, **89**, **97**, …

In fact, as you go higher the number of primes between squares seems to increase, apart from anything else because the space between squares increases, and yet you can demonstrate that this happens only for a finite number of squares. However high you go, you have to stop some time, and without a proof there's no *guarantee* that there aren't two successive square numbers without a prime between them.

Then there are the *gaps* between primes, which seem to fascinate mathematicians as much as the primes themselves. Martin Huxley is a slow-speaking, fast-thinking mathematician from the University of Cardiff.

'I'm told that when I was very small I used to read out the numbers of all the houses as we walked along the street,' he said. Now he's working on something a bit more complicated.

'The problem I most like is to do with the gaps between prime numbers. One of the interesting things is the maximum gap. How long does it take before you get a very long gap? We can't do it either way: we can't show there are infinitely many gaps of two [the Twin Primes problem], or even infinitely many gaps less than one-fifth of the average gap. I managed to show that there are infinitely many gaps less than 0.4 of the average gap.'

'Problems about primes have a very beautiful feature to them,' said Peter Sarnak at Princeton. 'It's not that we define some artificial things with a set of axioms and then ask, "Does this follow from these axioms?" The beauty of problems about primes is you can invent anything you want to solve them, so maybe someone comes along with a proof which is on entirely different lines – you can throw anything at this problem. You can throw any branch of mathematics, and many branches of mathematics were invented in part to solve related problems, so they are not entirely divorced from it.'

Yoichi Motohashi, of Nihon University, is one of the few Japanese mathematicians working in depth on the Riemann Hypothesis. '[Primes] are full of surprises and very mysterious,' he told me. 'They are like things you can touch, like that…', and he leant forward and tapped a low metal lampshade. 'In mathematics most things are abstract, but I have some feeling that I can touch the primes, as if they are made of a really physical material. To me, the integers as a whole are like physical particles.'

Henryk Iwaniec of Rutgers University, put matters very simply. 'I just *love* working in prime numbers,' he said, with passion in his voice.

One of the greatest number theorists of the twentieth century was G.H. Hardy, who spent much of his working life at the University of Cambridge. He wrote eloquently about his work and his own passion for number theory:

> The theory of numbers has always occupied a peculiar position among the purely mathematical sciences. It has the reputation of great difficulty and mystery among many who should be competent to judge; I suppose that there is no mathematical theory of which so many well qualified mathe-

maticians are so much afraid. At the same time it is unique among mathematical theories in its appeal to the uninstructed imagination and in its fascination for the amateur. It would hardly be possible in any other subject to write books like Landau's *Vorlesungen* or Dickson's *History*, six great volumes of overwhelming erudition, better than the football reports for light breakfast table reading.[6]

Hardy's estimate of what would appeal to the 'uninstructed imagination' may have been accurate for a small group of upper-middle-class, public-school-educated Englishmen, but today there can be few non-mathematicians who would choose Landau's *Vorlesungen* for 'light breakfast table reading' (although I suppose there are some who might if the alternative is reading 'the football reports').

But the history of the Riemann Hypothesis and the steps that led up to it give a very good indication of why some mathematicians are afraid of number theory. The subject is riddled with questions whose answers, though they appear to be attainable with a bit of hard work, turn out to present fiendish problems that are soluble only by venturing into entirely different fields of mathematics. The number theorist therefore has to be something of a jack of all trades.

The central question about the primes, from which the Riemann Hypothesis eventually emerged, has been asked for centuries if not for millennia. Mathematicians have always wondered how the primes are distributed among the entire sequence of whole numbers. Of the first numbers from 1 to 100, 25 per cent are primes. But of the numbers from 1 to 1,000, the proportion drops to about 17 per cent; from 1 to 10,000, it's 12.29 per cent; and so on. Except that for many years 'and so on' was one thing you couldn't say with any degree of certainty, since the precise relationship between the whole numbers and the primes was not even suspected until the brilliant insights of a young German genius were published in 1801.

Carl Friedrich Gauss, a German mathematician who flourished at the end of the eighteenth century and the beginning of the nineteenth, contributed to many fields of mathematics and was a great number theorist. His contribution to our understanding of primes was a major step forward on the path to the Riemann Hypothesis. Gauss's unusual abilities were noticed by his schoolteachers when he was seven. One day, his class was given the task of adding up the numbers from 1 to 100. Carl Friedrich had finished the task while his classmates were still struggling their way through the first few additions. This was not because he had a capacity for lightning addition, but because he noticed a short cut. He saw that you could pair the first 100 numbers as follows:

$$1 + 100, \ 2 + 99, \ 3 + 98, \ 4 + 97, \ 5 + 96, \ 6 + 95, \ 7 + 94,$$
$$8 + 93, \ 9 + 92, \ldots$$

He realized that this would make fifty pairs, each pair adding up to 101, making a total of 50×101. So the answer he arrived at long before any of his schoolmates was 5,050.

Gauss wrote a book called *Disquisitiones arithmeticae* which was published in the first year of the nineteenth century, in which he set out some key new ideas about number theory. One of the readers of this book was a certain 'Monsieur LeBlanc', who sent Gauss some of his own ideas before revealing himself later as a woman, one of history's rare women mathematicians, Sophie Germain. Gauss's letter to Germain after this revelation reveals a shared passion:

> But how can I describe my astonishment and admiration on seeing my esteemed correspondent Monsieur LeBlanc metamorphosed into this celebrated person, yielding a copy so brilliant it is hard to believe? The taste for the abstract sciences in general and, above all, for the mysteries of numbers, is very rare: this is not surprising, since the charms of this sublime science in all their beauty reveal themselves only to those who have the courage to fathom them. But when a woman, because of her sex, our customs and prejudices, encounters infinitely more obstacles than men, in familiarizing herself with their knotty problems, yet overcomes these fetters and penetrates that which is most hidden, she doubtless has the most noble courage, extraordinary talent, and superior genius. Nothing could prove to me in a more flattering and less equivocal way that the attractions of that science, which have added so much joy to my life, are not chimerical, than the favour with which you have honoured it.[7]

It was this taste of Gauss's for 'the mysteries of numbers' that led him to ponder the mystery of the primes, and in particular their distribution. Were they scattered randomly, like grains of salt on a tablecloth? Or was there some rule that told you how many primes there would be in a certain interval – from 1,000 to 5,000, say, or from 10^{23} to 10^{24}?

Since large numbers and how they are written features a lot in discussions of number theory, I should remind you that 10^{23} means 10 multiplied by itself 23 times. It is easy to work out, since when it is expanded it is a 1 followed by 23 zeros, but any number can be raised to any power, so 2^{100} means 2 multiplied by itself 100 times. In this case, of course, there's no easy way to expand it other than by doing the 99 sums – and getting a very big number.

Gauss's observations confirmed that as you went higher and higher in the whole numbers, the proportion of primes got less. This might sound a chore, but in fact Gauss told a friend that whenever he had a spare fifteen minutes he would spend it in counting the primes in a range of 1,000 numbers. By the end of his life it is estimated that he had counted all the primes up to about 3 million.

Samuel Patterson, a mathematician from Belfast who is now a professor at the University of Göttingen, explained what these data revealed. 'At the very simplest level you begin with something like what Gauss did – you count the number of primes, for example, between 1 and 1,000, between 1,000 and 2,000, and so on up to a million, or whatever Gauss actually worked out. You discover that these numbers are decreasing gently but very regularly, and then you start simply counting the number of primes up to 1,000, 2,000, 3,000. You plot them on a graph, and you find that the graph is very smooth where you don't expect it. You can just draw a straight line through it – it's very regular – you can see that it's a real phenomenon. The question is whether it has an illuminating proof or solution.'

Gauss saw that the number of primes in the interval from 1 to n, where n was any number you cared to choose, increased much more slowly than n. With the primes, when n went from 1 to 1,000 there were 168 primes in that interval, but the next 9,000 contained only another 1,061 primes, an average of 117 primes per 1,000. And by the time n reaches 10^{15}, there are only, on average, 30 primes per 1,000 whole numbers. Gauss also saw that this rate of change was characteristic of the way logarithms behave. The logarithms of the whole numbers increase much more slowly than the numbers themselves. The log of 1,000 is 3, the log of 10,000 is 4, the log of 100,000 is 5, and so on. (For more on logarithms and exponents, see *Toolkit 1*.)

If you didn't know better, you might think that as you explored higher and higher numbers the proportion of primes would eventually drop to zero. But, as we have seen, Euclid had knocked that idea on the head when he proved that there are an infinite number of primes. So the graph of the number of primes in each interval of a thousand, or a million, would be a line that slopes downwards but never reaches the horizontal axis, where the number would equal zero. Gauss thought about this and tried to guess the rule that governs how the primes are distributed. The answer he came up with was that it varied in a way that mathematicians call logarithmic.

Imagine you have a fourteen-year-old boy in the family. What might he be doing right now? Watching MTV? Reading a Harry Potter book? Kicking a football about? Painting graffiti on a wall? It's much more likely that he'd be engaged in one of these activities, rather than considering the distribution of prime numbers. But in the papers of Carl Friedrich Gauss is a copy of a table of logarithms that Gauss had obtained as a boy of fourteen. Then, as now, boys used to scribble on their schoolbooks, and Gauss's table of logarithms has on its back page some boyish graffiti:

Primzahlen unter a $(= \infty)$ a/la

This was the rule that Gauss believed governed the distribution of the prime numbers. It is what became known as the Prime Number Theorem. Put into words, it says that the number of prime numbers less than *a* as *a* approaches infinity (indicated by the symbol ∞) gets nearer and nearer to the value of *a* divided by the logarithm of *a*. So the number of primes less than a million is roughly equal to a million divided by the logarithm of a million.

The logarithm Gauss used is not a logarithm based on 10. If it were, we could say that the number of primes less than a million is about 166,666: 1,000,000/6, since 10^6 is a million. But it isn't 166,666 – it's 78,499, because the logarithms Gauss used were not based on powers of 10. Instead, he used a system based on a number that is important in all sorts of ways in mathematics, a number denoted by e, which has a value of 2.718 281 828.... But the system works in the same way. If $A = 10^L$, the logarithm of A (to the base 10) is L. Similarly, if $A = e^M$, the logarithm of A (to the base e) is M.

Gauss's observation of the behaviour of the distribution of prime numbers was profound. But it was only an observation. It was on a par with observing that however high a number you take, you can always find a prime beyond it. But that statement could not be taken as certain since it is limited by our human ability to count up to very high numbers. It was only when a proof became available to show a way of *constructing* a prime number larger than any known prime that we could be confident that there are an infinite number of primes.

In the same way, we couldn't *know* that Gauss's rule for the distribution of primes was a correct description until there was a proof. And that wasn't found until 1896, forty-one years after Gauss died. In that year, Jacques Hadamard, a French mathematician, published a proof that showed why the primes had to be distributed in the way Gauss suggested. And in the odd way that things sometimes happen with proofs (and buses), you wait decades for one to come along and then two come along together. Also in 1896, a Belgian mathematician, Charles de la Vallée-Poussin, published *his* proof of the Prime Number Theorem, which he'd worked out entirely independently of Hadamard.

The mathematical statement that the two men managed to prove was essentially Gauss's guess – that the number of primes less than or equal to a number x gets nearer to the value $x/\log_e x$ as x tends to infinity ($\log_e x$ means the logarithm of x to the base e; if it was to the base 10 it would be written $\log_{10} x$).

The reason for dealing with the Prime Number Theorem in some detail is because it forms a crucial link between the prime numbers and the Riemann Hypothesis. The link is in the phrase 'gets nearer to' used in the definition above, because this definition is not an exact statement. Each time you calculate the number of primes less than x for different values of x, you get an answer that is near the correct value but not exactly equal to it. (Obviously, for comparatively small values of x, we can actually count the primes so we can see how far off the correct answer the Prime Number Theorem is.) For example, if we count the primes less than 10,000,000,000, we find that there are exactly 455,052,512. The Prime Number Theorem predicts that there will be roughly 434,294,493 of them. An error of 20,758,019 would seem more likely get a schoolboy a detention than to attract the praise of the mathematical community. But the answer given by the Prime Number Theorem is out by only $4\frac{1}{2}$ per cent, and this is more accurate than anybody else had been able to achieve by the end of the nineteenth century.

At the time the Prime Number Theorem was proved, the Riemann Hypothesis had been around for about forty years. And it turned out that there was a link between the two mathematical statements. If the Riemann Hypothesis was true it could lead to an *exact* formulation of the Prime Number Theorem, instead of one that is always out by several per cent. You can rephrase the Prime Number Theorem to say that the number of primes less than x is $x/\log_e x$ plus some other number which is small in comparison. This is sometimes written as: the number of primes less than x is $x/\log_e x + \varepsilon$, where the Greek letter ε (epsilon) stands for the extra number – different for each x – that has to be added in to make the result accurate. Getting an exact figure for that ε is something which is promised by the Riemann Hypothesis – *if* it is true.

2 'Gorgeous stuff'

> I'm beginning to think that nothing is more conducive to the abstract sciences than prison. My Hindu friend Vij. often used to say that if he spent six months or a year in prison he would most certainly be able to prove the Riemann hypothesis. This may have been true, but he never got the chance.
>
> André Weil[8]

On 24 May 2000, the Clay Mathematics Institute of Cambridge, Massachusetts, one of whose aims is to 'further the beauty, power, and universality of mathematical thought', announced its prize for proving (or disproving) the Riemann Hypothesis, among a series of prizes of a million dollars for solving seven mathematical problems:

> Before consideration, a proposed solution must be published in a refereed mathematics journal of world-wide repute, and it must also have general acceptance in the mathematics community *two years* [after that publication]. Following this two-year waiting period, the SAB [the Scientific Advisory Board of the Institute] will decide whether a solution merits detailed consideration...The SAB will pay special attention to the question of whether a prize solution depends crucially on insights published prior to the solution under consideration. The SAB may (but need not) recommend recognition of such prior work in the prize citation, and it may (but need not) recommend the inclusion of the author of prior work in the award.

Pure maths is not usually associated with big money. But there have been a number of mathematical prizes awarded over the years for specific problems, or just for outstanding work. In September 2001, Professor Alain Connes, one of the leading mathematicians working on nn Hypothesis, was presented with a cheque for half a million m a Swedish organization, the Crafoord Foundation, simply for being a very good mathematician. In the case of the Riemann Hypothesis, it's difficult to see what effect the existence of the prize will have. Some mathematicians feel that the Clay Institute could have done better things with its money if it really wanted to increase the chances of a proof being discovered.

Professor Daniel Bump of Stanford University in California is one of them. 'If the Clay Institute wants to stimulate work on the Riemann Hypothesis, a far better use of a million dollars would be to fund projects rather than set up a prize and say, "Work on it for this reason." The Riemann Hypothesis doesn't need a million-dollar prize. If people have ideas about the Riemann Hypothesis they'll work on them without the [inducement of a] prize, so I don't think the reward is going to have any positive effect on research. What it does do is generate a certain amount of publicity, and so it may generate funding at a National Science Foundation level or similar. When Andrew Wiles proved Fermat's Last Theorem, that was good for mathematics – it generated a lot of publicity and it was on the front page of the *New York Times*. This is a way of getting the Riemann Hypothesis on the front page of the *New York Times*. It's not a way of stimulating research, because people don't need a million dollar prize to work on it or any of the problems they are offering prizes for.'

Professor Bump's office is in Building 380 on the campus of Stanford University. His website* indicated that I was likely to be meeting a mild eccentric. Among links to some of the current work on the Riemann Hypothesis was a link to pages about cats. The sketch of Bump on his homepage shows a hirsute individual surrounded by scribbled equations. In fact, that's exactly what he turned out to be. A fine two-pronged beard covered his chest, and as we talked he swiftly covered several pages of a yellow pad with helpful illustrations.

Unlike an artist or a musician, a mathematician can display the quality of his mind only to other like-minded people, through the maths he performs. And yet I could understand the deep nature of the Riemann Hypothesis only by trying to see into the minds of the people who were passionate about it, in the hope that I could share some of that passion. So with Bump, as with other mathematicians, I wanted to know how his interest in the topic had led him towards this Everest of mathematical problems.

* http://math.stanford.edu/~bump/

Daniel Bump's initiation into the delights of mathematics occurred when he came across an odd type of fraction called a continued fraction. 'I remember when I was in fifth or sixth grade I had a book which had a chapter on continued fractions, and I got quite excited about them. That was the earliest thing I can remember.'

Here's an example of a continued fraction:

$$\cfrac{1}{1+\cfrac{1}{1+1}}$$

It has a definite value which you can work out if you start at the bottom. Add 1 and 1 to make 2, and the lowest fraction becomes $1\frac{1}{2}$, and the whole fraction becomes 1 divided by $1\frac{1}{2}$, which is $\frac{2}{3}$. You can extend such fractions downwards to get more fractions, all of them involving only the number 1, and for each of them it's possible to work out a value, for example:

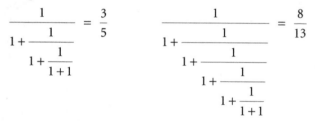

$$\cfrac{1}{1+\cfrac{1}{1+\cfrac{1}{1+1}}}=\frac{3}{5} \qquad \cfrac{1}{1+\cfrac{1}{1+\cfrac{1}{1+\cfrac{1}{1+1}}}}=\frac{8}{13}$$

The young Daniel Bump was intrigued when he was shown the following expression, which is 1 plus a continued fraction. In this case the fraction goes on for ever, signified by the familiar three dots.

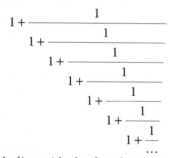

$$1+\cfrac{1}{1+\cfrac{1}{1+\cfrac{1}{1+\cfrac{1}{1+\cfrac{1}{1+\cfrac{1}{1+\cfrac{1}{\dots}}}}}}}$$

In dealing with the first few fractions we started at the bottom and worked up. But *this* fraction is bottomless. You can see how it goes on for ever. So does it have a value? Is there a specific number that we can say is the result of working out this continued fraction all the way to infinity, if that were possible? It turns out that an accurate value *can* be worked out. It's done by writing the expression as an equation and solving it. (For more about equations, see *Toolkit 2.*)

To find a value (call it *a*) for the continued fraction, we turn it into an equation that can then be solved with normal methods. The way we do this is based on the fact that the fraction below the first line is identical to the whole fraction. Normally a part of something is inevitably smaller than the whole, but here we have a part that is *equal* to the whole. This is the sort of thing that happens with infinity, an important concept in much of number theory and something that comes into the Riemann Hypothesis. If we call the value of the whole fraction *a*, then we have something like this:

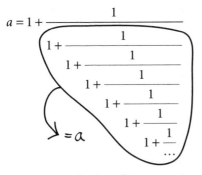

Take a moment to look at this expression. However you look at it, the part of it I've ringed is equal to the whole expression. The whole expression equals *a* – which is some number we are hoping to calculate – and so as long as we continue the levels of the fraction to infinity, we can write a new equation, a much simpler looking one:

$$a = 1 + \frac{1}{a}$$

Miraculously, most of those 1s have gone away and we have an equation that looks much more manageable than the continued fraction going off to infinity. This equation is a common type, a quadratic equation, and solving it means finding a number that will make a true statement when that number is substituted for *a*. That number turns out to be

$$a = \frac{\sqrt{5} - 1}{2}$$

which is therefore the value of that infinitely continued fraction we started with.

There was one other fact about this process that attracted the young Bump's attention. This number is called the Golden Mean, a number that crops up time and again in other areas of maths as well as in biology, botany, crystallography and even the visual arts.

'All these books about mathematics that they give children have stuff about the Golden Mean,' said Bump, 'and the mystical properties of it and so forth, and so that was the first mathematical thing that I can remember that stimulated my interest in mathematics at age of eleven.'

Throughout this interesting exposition, Bump never smiled. Maths was clearly a serious business for Professor Bump, and the mildly humorous remarks I made occasionally as we talked hung in the air, unacknowledged. The only thing that mattered to him was mathematics; and the thought of earning a million dollars was far from his mind.

Other mathematicians agree. 'The real mathematician would go for it anyway,' said Martin Huxley, 'whether or not it was somebody's prize problem.'

For Roger Heath-Brown, in Oxford, it's competing with his colleagues that matters. 'I'm afraid I see all of mathematics as a competition. I'm a very competitive person. I always want to prove a theorem for myself. If someone has a theorem, I always want to prove a better theorem.'

When I went to visit Louis de Branges, he told me why he'd decided not to accept the prize if he ever won it. 'I believe there needs to be a mathematical institute in Bourcia [the French village from which de Branges's ancestors came], and the needs of this institute [should] take priority over my needs, and therefore I would not accept the prize...I also think that if I prove the Riemann Hypothesis I would be protected by society; that my university would take a different view of me, they would keep me longer, and that I would get benefits that would be more than the benefit of a million dollars. If I had a million dollars I would first of all be heavily taxed; I would be seen as somebody that would be working for the million dollars; there would be perhaps relatives of mine that would require help, and what benefit would that bring to my work, to have a million dollars?' He looked around his apartment. 'I could perhaps improve this apartment or my house, so that there would be a first-class bathroom, or first-class kitchen. Sure, that would be nice, but it isn't my objective.'

I was to observe much evidence of De Branges's otherworldliness during the times I was with him, but as a mathematician who thinks only about mathematics, sometimes at the expense of his daily life, he is not unique. Without wishing to put him in the same category – that has to await de Branges's proof of the Riemann Hypothesis – Isaac Newton showed some of the same qualities:

> On one occasion, when [Newton] was giving a dinner to some friends at the university, he left the table to get them a bottle of wine; but, on his way to the cellar, he fell into reflection, forgot his errand and his company, went to his chamber, put on his surplice, and proceeded to the chapel. Sometimes he would go into the street half dressed, and on discovering his condition, run back in great haste, much abashed. Often, while strolling in his garden, he would suddenly stop, and then run rapidly to his room, and begin to write, standing, on the first piece of paper that presented itself. Intending to dine in the public hall, he would go out in a brown study, take the wrong turn, walk a while, and then return to his room, having totally forgotten the dinner. Once having dismounted from his horse to lead him up a hill, the horse slipped his head out of the bridle; but Newton oblivious, never discovered it till, on reaching a tollgate at the top of the hill, he turned to remount and perceived that the bridle which he held in his hand had no horse attached to it.[9]

For some present-day number theorists, the Riemann Hypothesis has become as much of a preoccupation as any of the problems that led Newton to miss his dinner or forget his horse. It is simply the most significant topic around.

'This is a first-class problem', Matti Jutila, a Finnish mathematician, told me. 'Compared, for instance, with Fermat's Last Theorem, I have always felt that Riemann's Hypothesis is genuinely important.'

Hugh Montgomery told me that within mathematics the Riemann Hypothesis is the problem most pure mathematicians would like to see solved. 'It's the one that many people have worked on, that is known to have many consequences, and it lies at the core of a famous old area, namely the study of prime numbers. So if you could be the Devil and offer a mathematician to sell his soul for the proof of one theorem – what theorem would most mathematicians ask for? I think it would be the Riemann Hypothesis.'

Henryk Iwaniec is a Polish-American mathematician whose eyes gleam when he discusses the Riemann Hypothesis. 'I would trade everything I know in mathematics for just knowing the proof of the Riemann Hypothesis. It's just gorgeous stuff. I'm only worried that what may happen is that a proof will be given by somebody and I will be unable to understand it. That's the worst…'

Iwaniec shows an intensity of enthusiasm that someone else might display for a favourite piece of music, or a rare and delicious meal. 'The consequences are fantastic: the distribution of primes, these elementary objects of arithmetic. And to have tools to study the distribution of these objects – that's the discovery of Riemann. There's a whole lot of things combined. Great scholars have tried it; your own perception that this is fundamental; the beauty and difficulty; and connections with so many areas. I think it is right to be named as the number one unsolved problem in mathematics.'

Peter Sarnak explained to me how much rides on a solution. 'If it's not true, then the world is a very different place. The whole structure of integers and prime numbers would be very different to what we could imagine. In a way, it would be more interesting if it were false, but it would be a disaster because we've built so much round assuming its truth.'

So far, we've got a sense from working mathematicians that the Riemann Hypothesis says something fundamental about prime numbers. To recap: a formula devised by Gauss which purports to calculate the number of prime numbers up to any number x (the Prime Number Theorem) is not quite accurate. If you actually count the number of primes less than any number and compare the result you get by using the formula, there is always a difference of a few per cent. So you have the predicted number plus an error. The predicted number is known and was proven in 1896, by the Prime Number Theorem. And the value of the error is what the Riemann Hypothesis provides – if it's true.

Bernhard Riemann was a pupil of Gauss's, so he would have known all about Gauss's guess for the number of primes in a particular interval. He would also have thought that, nice as it is to have any figure at all for the number of primes less than n, it would be nicer to have a better estimate. Even when n was as large as 1,000,000,000,000,000, Gauss's formula gave a result that was out by about 3 per cent.

Riemann came up with another formula which he thought would

give a more accurate result. Let's call it RF(n). Now, when you substituted 1,000,000,000,000,000 for n, the answer Riemann's formula gave for the number of primes less than 1,000,000,000,000,000 was still inaccurate, but by far less than 3 per cent. In fact, the error was just

$$\frac{245}{100,000,000,000} \text{ per cent}$$

But Riemann still wasn't content. He didn't want any inaccuracy at all, even if it were less than a hundred millionth of 1 per cent. So he added one final factor to his formula, the sum of an infinite series, represented in mathematical notation as

$$\sum \frac{1}{n^s}$$

Now, in the words of Douglas Adams, 'Don't panic!' Mathematical symbols can be very off-putting. It sometimes seems that they embody a secret known only to mathematicians.

This extract (Figure 2) from *Principia Mathematica* by Bertrand Russell and Alfred North Whitehead is absolutely impenetrable to the non-mathematician. In fact, as you'll see when you read the concluding sentence, it is a proof that $1 + 1 = 2$. Whitehead wrote an eloquent

---

∗54·43. $\vdash :. \alpha, \beta \epsilon 1 . \supset : \alpha \cap \beta = \Lambda . \equiv . \alpha \cup \beta \epsilon 2$

Dem.

$\vdash . \ast 54 \cdot 26 . \supset \vdash :. \alpha = \iota^{\prime} x . \beta = \iota^{\prime} y . \supset : \alpha \cup \beta \epsilon 2 . \equiv . x \neq y .$

[∗51·231] $\equiv . \iota^{\prime} x \cap \iota^{\prime} y = \Lambda .$

[∗13·12] $\equiv . \alpha \cap \beta = \Lambda$ (1)

$\vdash . (1) . \ast 11 \cdot 11 \cdot 35 . \supset$

$\vdash :. (\exists x, y) . \alpha = \iota^{\prime} x . \beta = \iota^{\prime} y . \supset : \alpha \cup \beta \epsilon 2 . \equiv . \alpha \cap \beta = \Lambda$ (2)

$\vdash . (2) . \ast 11 \cdot 54 . \ast 52 \cdot 1 . \supset \vdash . \text{Prop}$

From this proposition it will follow, when arithmetical addition has been defined, that $1 + 1 = 2$.

FIGURE 2 Conclusive proof that $1 + 1 = 2$, from Russell and Whitehead's *Principia Mathematica*.

defence of the use of symbols in mathematics:

> Mathematics is often considered a difficult and mysterious science, because of the numerous symbols which it employs. Of course, nothing is more incomprehensible than a symbolism which we do not understand. Also a symbolism, which we only partially understand and are unaccustomed to use, is difficult to follow. In exactly the same way the technical terms of any profession or trade are incomprehensible to those who have never been trained to use them. But this is not because they are difficult in themselves. On the contrary they have invariably been introduced to make things easy. So in mathematics, granted that we are giving any serious attention to mathematical ideas, the symbolism is invariably an immense simplification.[10]

Some mathematical symbolism is easy to understand because it uses symbols that are more or less part of everyday life: $+, -, \times, \div, =, >, <, (\ )$. But even these symbols are capable of subtleties that are more important in mathematics than in ordinary usage. In some areas of mathematics, $a + b$ does not necessarily equal $b + a$. That's one of the reasons why Russell and Whitehead went to such lengths to prove that $1 + 1 = 2$. *Nothing* can be taken for granted. (Even in everyday life, when giving directions to someone, 'turn left then turn right' doesn't always gets you to the same place as 'turn right then turn left'.) The British physicist Sir Arthur Eddington once said, 'We used to think that if we knew one, we knew two, because one and one are two. We are finding that we must learn a great deal more about "and".'[11]

Bertrand Russell had a bit of fun instilling doubt into someone who believed that $2 + 2$ always equals 4:

> You are quite right, except in marginal cases – and it is only in marginal cases that you are doubtful whether a certain animal is a dog or a certain length is less than a meter. Two must be two of something, and the proposition '2 and 2 are 4' is useless unless it can be applied. Two dogs and two dogs are certainly four dogs, but cases arise in which you are doubtful whether two of them are dogs. 'Well, at any rate there are four animals,' you may say. But there are microorganisms concerning which it is doubtful whether they are animals or plants. 'Well, then living organisms,' you say. But there are things of which it is doubtful whether they are living organisms or not. You will be driven into saying: 'Two entities and two

entities are four entities.' When you have told me what you mean by 'entity,' we will resume the argument.[12]

Even Goethe had a view on the topic: 'Two times two is not four, but it is just two times two, and that is what we call four for short. But four is nothing new at all.'[13]

For a Mrs La Touche, a Victorian correspondent to the *Mathematical Gazette, nothing* was certain when it came to addition:

> There is no greater mistake than to call arithmetic an exact science. There are permutations and aberrations discernible to minds entirely noble like mine; subtle variations which ordinary accountants fail to discover; hidden laws of number which it requires a mind like mine to perceive. For instance, if you add a sum from the bottom up, and then from the top down, the result is always different.[14]

Mrs La Touche was maybe more perceptive than she realized, since there are some mathematical systems where the order of addition affects the end result. But not with the series of additions that are at the heart of Riemann's correction factor for the number of primes less than n. (For more about infinite series, see *Toolkit 3*.)

To recap the symbols that describe the infinite series that is important to the Riemann Hypothesis: in $\Sigma 1/n^s$, the symbol Σ (a large Greek capital sigma) actually means 'add together whatever follows it', so $\Sigma 1/n^s$ means 'Generate a series of terms of the form $1/n^s$, with s some fixed number and n increasing by 1 in each term, and add all the terms together.' You might think that adding an infinite series of terms together will produce an infinitely large result, but this is not always the case. It depends in this instance on the value of s. Such a process, with the right choice of s, can give a result that is a fixed number.

So, calculating the value of the sum $\Sigma 1/n^s$ (call it S), which Riemann believed was possible but couldn't say so for certain, would result in a totally accurate number for the number of primes less than n.

To summarize so far:

What is the total number of primes less than any number n?
Gauss's guess: $n/\log n$ – out by several per cent.
Riemann's first guess: $RF(n)$ – out by a fraction of 1 per cent.
Riemann's better guess: $RF(n)$ minus the sum of the infinite series S –
BULLSEYE!

This number *S*, the sum of the infinite series $\Sigma\,1/n^s$, is what is called the zeta function (or ζ function, using the Greek letter zeta), and it was originally investigated by a mathematician called Leonhard Euler* in the 1730s.

Like many of the giants of mathematics, Euler lived for mathematics, so much so that upon the loss of his right eye he said, 'Now I will have less distraction.'[15] Among his many feats, some of them non-mathematical, was his ability to 'repeat the *Æneid* from the beginning to the end, and he could even tell the first and last lines in every page of the edition which he used. In one of his works there is a learned memoir on a question in mechanics, of which, as he himself informs us, a verse of *Æneid* gave him the first idea.'[16] The verse contained the line, 'The anchor drops, the rushing keel is staid.'

Although Euler made one of the most important contributions to the study of prime numbers, he had no great confidence that we would ever understand them, so he would probably not have been surprised to learn of the problems mathematicians are having more than two hundred years after his death. 'Mathematicians have tried in vain to this day to discover some order in the sequence of prime numbers,' Euler wrote, 'and we have reason to believe that it is a mystery into which the human mind will never penetrate.'[17]

Euler's zeta function, as we've seen, is a sum of fractions involving the whole numbers – counting from 1 to infinity – and the letter *s*. A function is another of those words that has one use in everyday life and a very different and much more specific one in mathematics. You can think of it as a black box, with an input and an output. Whatever you input is transformed in some way and emerges as a different output. So if you put in the succession of numbers 1, 2, 3, 4, 5, 6, and the function turned them into 1, 4, 9, 16, 25, 36, you would call the function x^2. Functions are everywhere in mathematics, and they all work by putting some number in at one end and seeing what comes out at the other. But it's the relationship between the *succession* of inputs and outputs, rather than individual pairings of numbers, which the mathematicians are really interested in. They talk about the 'behaviour' of a function: does it behave in a regular way, with a steady increase in output as the input increases or decreases? Does it suddenly change its behaviour at a

---

* Pronounced 'oiler'.

certain point and go off to infinity or plummet to zero? Are there some values of input for which there is no output at all? And so on.

The mathematics of functions is called analysis, and one branch of analysis is very useful for looking at the whole numbers. It's called analytic number theory. Charles Ryavec came up with a graphic illustration to help me understand this process:

'I was on a trip once,' he said, 'and we went through Yellowstone Park. There was this ranger and he was giving us a tour, and he told us that once you could come here to a geyser, and he could take your handkerchief and throw it into the water here and it would suck it down, and about forty minutes later it came out over there all washed for you. This is basically analytic number theory. But then he said that someone had come with a whole basket of laundry and the geyser was clogged up now and they had no way of unclogging the whole thing, and that also is analytic number theory!' Ryavec put back his head and laughed. 'In a funny way, you take numerical information and you stick it into a function and then somehow the function produces more than what you put in, or it sort of reassembles or cleans up what you put in and gets nice results.'

Euler discovered a relationship between his zeta function, which uses all the whole numbers, and an entirely different series which uses only the prime numbers. We've already seen how unpredictable the prime numbers are: they don't occur regularly in the long line of whole numbers; and there can be long stretches along the road of real numbers where there are no primes, then two (but not three) come along almost together. Nevertheless, Euler made what one mathematician has described as 'one of the most remarkable discoveries in mathematics'. He discovered that the zeta function can be expressed as a series of terms in which *adding* all the *whole* numbers was exactly equivalent to another function which consists of *multiplying* together a series of terms involving all the *prime* numbers. So feeding the same value for s into each of these two very different functions produced the same result.

It's difficult to think of an exact analogy that conveys what is remarkable about this, but it is rather as though – as I used to think as a child – you could translate English into French by an alphabetic code. If the code had $c=n$, $l=u$, $o=a$, $u=g$ and $d=e$, and you tested it with the English word 'cloud', sure enough you get the French word 'nuage'. It would be extraordinary if that worked for any other examples ('old' is not 'ane' in French), but it is truly amazing that Euler found an equivalent way of translating a series about the whole numbers into another series about primes.

His discovery was that:

$$\zeta(s) = \sum \frac{1}{n^s} = \prod \frac{p^s}{p^s - 1}$$

Remember, Σ (a large Greek capital sigma) means the *sum* of the terms in a series. In this equation there is also Π (a large Greek capital pi), which means the *product* of all the terms in a series, the answer you get by multiplying them all together.

Taking it step by step, Euler's discovery says that if you *add* together the reciprocals of all the whole numbers squared, for example, taking

$$s = \frac{1}{1^2} + \frac{1}{2^2} + \frac{1}{3^2} + \frac{1}{4^2} + \frac{1}{5^2} + \cdots$$

will give the same answer as *multiplying* together a series of terms involving the primes, each of the form $(p^2/p^2 - 1)$. So replacing each p with the prime numbers in order, 2, 3, 5, 7, 11, ..., the first term in this new series is $2^2/(2^2 - 1) = 4/3$, the second term is $3^2/(3^2 - 1) = 9/8$, the next term is $5^2/(5^2 - 1) = 25/24$, the next is $7^2/(7^2 - 1) = 49/48$, and so on through all the prime numbers to infinity. Then these terms have to be multiplied together:

$$\frac{4}{3} \times \frac{9}{8} \times \frac{25}{24} \times \frac{49}{48} \times \cdots \times \infty$$

Now, the fact that these two apparently different series are actually equivalent is not just an interesting coincidence (like the French 'nuage–cloud' example I manufactured). It was clear to Euler that it is a fundamental relationship between the whole numbers and the prime numbers, just the sort of thing that mathematicians have looked for as a means of understanding the whole field of numbers. (For a proof of the identity between these two series, see *Toolbox 4*.)

But this relationship doesn't actually get you very far if you want some specific knowledge about, say, the nth prime number, or the prime number that comes after a specific prime you know already. It was this next step that Riemann was after, and he thought he would get nearer his goal with his new function, the Riemann zeta function, which is a more sophisticated version of Euler's zeta function.

So let's get back to Riemann's formula, which I simplified as 'RF(n) minus the infinite series S'. This last term, the sum of the series S that corrects for the discrepancy between Gauss's guess and the correct answer, uses Riemann's version of Euler's zeta function. Riemann devised his own function, with a rather special value of the exponent s (not to be confused with S, the sum of the series), and it is now called the Riemann zeta function. This means that as you use Riemann's formula to calculate the number of primes in a particular interval – the 'density' of the primes – the answers vary in some mysterious way which is related to the behaviour of Riemann's zeta function. The behaviour of that zeta function is nothing like the behaviour of Euler's zeta function.

As we can see, Euler's zeta function varies in a regular way as s increases. For example, the value of $\zeta(2)$, which is written out as

$$\frac{1}{1^2} + \frac{1}{2^2} + \frac{1}{3^2} + \frac{1}{4^2} + \frac{1}{5^2} + \cdots$$

turns out to equal $\pi^2/6$, about $1.644\,934\,43$. As s increases, the value of $\zeta(s)$ decreases, so $\zeta(4) = 1.0823$. But *Riemann's* zeta function behaves in a very different way, fluctuating wildly and even sometimes equalling zero for some values of s. Those values of s that make Riemann's zeta function zero hold the secret of how the primes are distributed. That's because s in the Riemann zeta function is a different type of number altogether from the s in Euler's zeta function. Riemann used numbers which mathematicians call 'imaginary numbers'. And that's a whole new story.

3 New numbers for old

> The imaginary number is a fine and wonderful recourse of the divine spirit, almost an amphibian between being and not being.
>
> Gottfried Wilhelm Leibniz

It's very surprising that imaginary numbers are so unknown outside the field of mathematics. You don't have to delve very deeply into mathematics to come across these numbers 'whose nature is very strange but whose usefulness is not to be despised', as Leibnitz said. They are also at the heart of many applications of mathematics, from physics and electronics to astronomy and astrophysics. In the story of the Riemann Hypothesis, imaginary numbers, part of a number system called complex numbers, are so fundamental that we can't get much further without them.

Imaginary numbers emerged from a process that has often occurred in the history of mathematics, when mathematicians start off inside the borders of a particular mathematical system and then ask questions about what happens on the outside. This 'what if' approach can lead to a continual injection of new concepts as mathematicians realize that they have outgrown the old ones. What sometimes makes these innovations so difficult for non-mathematicians to understand is the popular belief that the concepts of mathematics are in some way rooted in everyday life. This impression was reinforced for many of us at school by the problems we were given to solve, involving baths filling with water, flagpoles casting shadows, men digging trenches, or people whose fathers were half as old fifteen years ago as they will be in ten years' time. Numbers in arithmetic and lengths in geometry all had a reassuringly familiar feel to them. Even algebra, with its more mysterious xs, could be seen as a method of finding out what change you had left after you'd bought seven pies, three of which cost twice as much as the other four.

But the familiarity is deceptive. Mathematicians delight in disputing the familiar by questioning its narrow boundaries. If we had discovered that the amount of water left in the bath was the square root of minus one, or if your father's age was negative five years ago, or the parallel sides of the square we had constructed actually met, we'd have felt sure that we'd missed out a crucial step in the equation. But for mathematicians nothing is unimaginable, and if the correct steps in reasoning lead to an unfamiliar or counter-intuitive answer, they sometimes see that as the starting point for a new journey.

In the history of mathematics there have been many occasions when, in order to advance the subject, mathematicians have had to create concepts that might on the face of it seem to make no sense. There was a time, for example, when the idea of a negative number was abhorrent. For the Greeks, numbers were chunky, *positive* things, usually associated with very visible objects. The Greek mathematician Diophantus, who flourished around AD 250, devised a type of equation which is now called Diophantine, with solutions in whole numbers, and he wouldn't accept any solutions to those equations that were numbers we would now call negative. So Diophantus and his disciples would consider an equation such as $x^2 - 45x = 250$ to have only one solution, $x = 50$ (because $50^2 - 50 \times 45 = 250$). But in fact, when Brahmagupta, an Indian mathematician who lived four hundred years after Diophantus, decided that negative numbers were valid solutions to equations, it became clear that the same equation, $x^2 - 45x = 250$, would work if $x = -5$. This is because $(-5)^2 = (-5) \times (-5) = +25$, and $-45 \times -5 = 225$ (because two negative numbers multiplied together produce a positive result), and $25 + 225 = 250$.

The first successful attempts to use negative numbers were met with disbelief since no one could imagine a negative number of pebbles or apples or cows. But even when people couldn't quite work out what they meant in the real world, negative numbers turned out to make some calculations much more consistent. The introduction of these new and somewhat ethereal numbers meant that *all* quadratic equations (equations involving x and x^2, but no higher powers of x) now had two solutions, whereas previously, when only positive numbers were deemed acceptable, they could sometimes have one, sometimes two, sometimes none.

Although early mathematicians knew about quadratic equations, they often didn't know that they always have two roots – that is, two answers that will work when substituted for x. Equations called linear equations, such as $x-3=0$, have only one answer. In this case, $x=3$. But $x^2-9=0$ has two answers: $x=3$ and $x=-3$. And with some quadratic equations the answers aren't always the same figure with different signs. There are two possible values of x for which $x^2+x-12=0$. These are $+3$ and -4. Because the earliest mathematicians lacked any concept of negative numbers, they found that some quadratic equations had one root and others two. To a good mathematician, this oddity might have seemed worth exploring and could have led to the use of negative numbers.

Even today, although we think we understand the concept of negative numbers, they can still prove a stumbling block to schoolchildren who are taught rules to manipulate them without being given the chance to understand why those rules work. How many of us have echoing in the dim recesses of our brains the phrase 'a minus times a minus is a plus'?

For Ruth McNeill, this 'rule' drove her to abandon maths:

What did me in was the idea that a negative number times a negative number comes out to a positive number. This seemed (and still seems) inherently unlikely – counterintuitive, as mathematicians say. I wrestled with the idea for what I imagine to be several weeks, trying to get a sensible explanation from my teacher, my classmates, my parents, anybody. Whatever explanations they offered could not overcome my strong sense that multiplying intensifies something, and thus two negative numbers multiplied together should properly produce a very negative result. I have since been offered a moderately convincing explanation that features a film of a swimming pool being drained that gets run backwards through the projector. At the time, however, nothing convinced me. The most common-sense of all school subjects had abandoned common sense; I was indignant and baffled. Meanwhile, the curriculum kept rolling on, and I could see that I couldn't stay behind, stuck on negative times negative. I would have to pay attention to the next topic, and the only practical course open to me was to pretend to agree that negative times negative equals positive. The book and the teacher and the general consensus of the algebra survivors of society were clearly more powerful than I was. I capitulated. I did the rest of algebra, and geometry, and trigonometry; I did them in the advanced sections, and I often had that nice sense of 'aha!' when I could suddenly see how a proof was going to come out.

Underneath, however, a kind of resentment and betrayal lurked, and I was not surprised or dismayed by any further foolishness my maths teachers had up their sleeves... Intellectually, I was disengaged, and when maths was no longer required, I took German instead.[18]

In fact, for those people in the same boat as Ms McNeill, Israel Gelfand, a brilliant mathematician who takes a strong interest in maths education, gave the following succinct and possibly convincing explanation:

3 times 5 = 15:
   Getting five dollars three times is getting fifteen dollars.

3 times (−5) = −15:
   Paying a five-dollar penalty three times is a fifteen-dollar penalty.

(−3) times 5 = −15:
   Not getting five dollars three times is not getting fifteen dollars.

(−3) times (−5) = 15:
   Not paying a five-dollar penalty three times is getting fifteen dollars.[19]

One of the most shocking mathematical innovations – at the time, during the second half of the seventeenth century – was introduced by Isaac Newton and Gottfried Wilhelm Leibniz when they independently devised a new field of mathematics which became the calculus. They were attempting to deal with *continuity* in mathematics in an era when much of mathematics dealt with discrete numbers. Their work involved curves, which can be thought of as being made up of points which could be as close to each other as you like. If you imagine two separate points on a curve joined by a straight line and moving towards each other, then when the points meet, the line joining them becomes a tangent to the curve, touching it at one point, the meeting point, instead of cutting the curve at two points.

To find out more about the tangent, Newton devised something called an infinitesimal. As the points approach each other, the distance between them gets smaller and – very roughly – the infinitesimal could be seen as the distance between the points at the instant they touched. He was able to use this 'invention' to find the slope of the tangent.

Bishop Berkeley, in the first half of the eighteenth century, attacked the idea of infinitesimals, saying that they were just as unscientific as some of the ideas held by him and his fellow theologians, for which they had been mocked by scientists. He wrote a polemic called *The Analyst. Or, a discourse addressed to an infidel mathematician. wherein it is examined whether the object, principles, and inferences of the modern analysis are more distinctly conceived, or more evidently deduced, than religious mysteries and points of faith. 'first cast the beam out of thine own eye; and then shalt thou see clearly to cast out the mote out of thy brother's eye.* After ridiculing the ideas of the calculus, Berkeley continued thus:

> All these points, I say, are supposed and believed by certain rigorous exactors of evidence in religion, men who pretend to believe no further than they can see. That men who have been conversant only about clear points should with difficulty admit obscure ones might not seem altogether unaccountable. But he who can digest [infinitesimals] need not, methinks, be squeamish about any point of divinity.[20]

The bishop would have hated the next addition to the pantheon of strange numbers, which already included negative numbers and infinitesimals, when mathematicians began to use an even stranger class of numbers called complex numbers.

Mathematicians call the familiar numbers we all know the real numbers, and the set of all real numbers is called R. You can think of these as lying on a straight line that extends either side of 0, stretching to infinity in one direction and to minus infinity in the other. It's a version of the street we visited in Chapter 1 where there were houses at intervals inhabited by members of the different prime families. As we look along this street of our numbered 'houses' – primes and composites – it looks rather empty. It certainly doesn't, at the moment, represent all the numbers we can think of, just the whole ones. But what about, say, fractions? You might think of building a house numbered $3\frac{1}{2}$ – people sometimes do that on a spare plot of land between two houses – but there certainly isn't room for $3\frac{1}{3}$, $3\frac{1}{4}$, $3\frac{1}{5}$, $3\frac{1}{6}$ or $3\frac{1}{7}$, nor could you fit in $3\frac{83}{111}$ or, indeed, any of the infinite number of other fractions you could dream up that lie between 3 and 4. For the time being, let's just say they're represented by a series of marks on the roadside, even though it's no easier to fit in an infinite number of marks than it is to fit in an infinite number of houses. But the essential point is that the fractions and the whole numbers all lie along the same straight line – the street.

Now for a leap into the unknown. It so happens that mathematicians have created a type of number that is *not* to be found on the street of real numbers. These numbers, called complex numbers, spread out on either side of that street, into the unurbanized landscape that we have so far thought of as bare and uninteresting. Without yet worrying about how or why the mathematicians have done this, let's just say that these numbers run along streets that are at right angles to the main street of real numbers. So if you go along Real Street – perhaps we'll call it Camino Real* – until you get to, say, number 10, and then turn left, you'll find yourself walking along another street. Here there are no houses, just addresses. You'll find numbers 1, 2 and 3, as well as $17\frac{2}{3}$, 455.3 and 100,000. There are other roads running off Camino Real all the way along it, each with many addresses stretching away to infinity. And, of course, many of the addresses might seem the same. On every street, including Camino Real, there's a number 27, a number 1,729, a number 3,000,000. If we want to identify a particular point, as a site for a new house for example, how can we give it a unique address? The usual way in the real world is to give names to the streets. But since it turns out that there are an infinite number of streets in this urban landscape, we'd have to find an infinite number of names.

Mathematicians have come up with a better idea: each address has two parts. The first part is the point on Camino Real where the side street begins; the second part, prefaced with an 'i' for 'imaginary', is the address along the side street itself. So (55 + 27i) is the twenty-seventh whole number along the street that meets the main road at house 55.

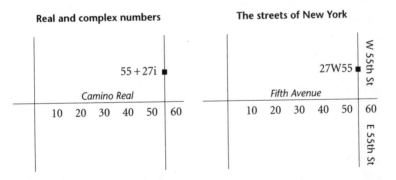

FIGURE 3 The system of representing complex numbers (left) is analogous to the addresses in New York streets (right).

---

* In fact, 'Camino real' in Spanish means royal road rather than real road, but it's a useful mnemonic.

If the street crosses Camino Real and continues on the other side, the numbering would have a minus, so an equivalent house on the lower part of the side street would be $55 - 27i$. It's not very different from the way buildings are numbered in a city such as New York. The address of the 26th building on West 55th Street is 26 W55, where 'W' means west of 5th Avenue, and there's a building on the continuation of 55th Street on the other side of Fifth Avenue called 27 E55 St. So Fifth Avenue and the side streets of New York can be seen as an analogue to the real and complex number field (see Figure 3).

These numbers can be treated in the same way as real numbers, so long as you keep all the i s together. You can add them together, for example, so that $(a+ib)+(c+id) = a+ib+c+id = a+c+i(b+d)$. And you can multiply them, so that $(a+ib) \times (c+id) = ac+iad+ibc+i^2bd$. This can be rewritten as $ac+i(ad+bc)+i^2bd$.

The i is a useful marker which enables us to identify the part of the complex number that is on an axis at right angles to the real number axis. But it turns out to have a much more useful function in certain types of equations: i is a number which can be taken to represent the concept 'the square root of -1'.

It's rare that literature deals with mathematics in an informed way. But there's a conversation in the novel *Young Törless*, by the German writer Robert Musil, that illustrates the puzzling nature of i. Two boys, Törless and Beineberg, have just come out of a maths lesson. Törless speaks first:

'I say, did you really understand all that stuff?'

'What stuff?'

'All that about imaginary numbers.'

'Yes. It's not particularly difficult, is it? All you have to do is remember that the square root of minus one is the basic unit you work with.'

'But that's just it. I mean, there's no such thing. The square of every number, whether it's positive or negative, produces a positive quantity. So there *can't* be any real number that could be the square root of a minus quantity.'

'Quite so. But why shouldn't one try to perform the operation of working out the square root of a minus quantity, all the same? Of course it can't produce any real value, and so that's why one calls the result an imaginary one. It's as though one were to say: someone always used to sit here, so let's put a chair ready for him today too, and even if he has died in the meantime, we shall go on behaving as if he were coming.'

'But how can you when you know with certainty, with mathematical certainty, that it's impossible?'

'Well, you just go on behaving as if it weren't so, in spite of everything. It'll probably produce some sort of result. And after all, where is this so different from irrational numbers – division that is never finished, a fraction of which the value will never, never, never be finally arrived at, no matter how long you may go on calculating away at it? And what can you imagine from being told that parallel lines intersect at infinity? It seems to me if one were to be over-conscientious there wouldn't be any such thing as mathematics at all.'[21]

At first sight it seems unlikely that any negative number could have a square root. According to the multiplication rules, including 'minus times minus is plus', there is no number that, when multiplied by itself, results in a negative number. Plus times plus is plus; minus times minus is plus. So how can there be a square root of *minus* 1? How can there be a number x, such that $x^2 = -1$? Well, as with the negative numbers, having a symbol to represent the square root of -1 turned out to have some use in solving equations.

Here's a quadratic equation, a deceptively simple one, that raises the problem:

$$x^2 + 1 = 0$$

It doesn't take too long to work out that the number that will solve this equation when substituted for x is a number which, when multiplied by itself, will equal -1. Then the equation would be transformed into $-1 + 1 = 0$, which is a true if not terribly interesting statement. But there is no 'number' in the sense that there is one of those digits we are all familiar with whose square equals -1. However, because it would be so useful if there were (in ways that will become apparent), mathematicians invented one. And as with π and e, they gave it a letter for a name and called it i.

In one sense, there's no such number – at least not in the sense that we use the word 'number' in everyday life. And in some ways it feels a bit of a cheat. Initially, it's quite intriguing to know that there *might* be a number which, when squared, equals −1. You could imagine being asked the question in a quiz, sitting down, and thinking very hard until, finally, you give up. 'I give up,' you say to the quizmaster, 'what's the answer?' When he then says, 'It's a symbol I've devised, called i', you could be forgiven for feeling a little disappointed, as if he's asked you what sort of circle has four corners, and then told you that the answer is a square circle.

But there's one crucial difference. Once i was invented, there were all sorts of situations in which it really *was* the square root of −1 in quite a profound sense. Certainly, as the idea of i began to catch on, the more sophisticated mathematicians began to see the immense value of these imaginary numbers. Using the map of Camino Real and its infinite side streets, mathematicians have devised a way in which using i^2 to represent −1 makes it possible to use complex numbers in the same way as real ones − multiplying them to get a meaningful answer, as well as adding them.

When mathematicians create new types of object, they like to devise or discover the rules that govern relationships between those objects. Just as there are rules that govern the relationships between the real numbers − the rules of addition, subtraction, multiplication and division − any new numbers, such as complex numbers, which are outside the real numbers should have their own set of rules. One of the things that happens with the real numbers is that however you apply the four operations of addition, subtraction, multiplication and division to individual numbers, the result is also a real number. 2 plus 3 times 17 divided by 8 is a real number. This may sound obvious, but it's not true of all classes of numbers, even simple ones. Take the integers, for example − the whole numbers. Addition, subtraction and multiplication of integers always produces more integers. But division doesn't. Sometimes it produces fractions: 12 divided by 3 produces another integer, but 11 divided by 3 doesn't.

All this is leading up to showing how, if we interpret i^2 to mean −1 in our new system of complex numbers, we can multiply two complex numbers together to produce a third that is also part of the same system. In other words, if we multiply two addresses together, we get a third address somewhere else on the street map. We can look at two addresses, say $(3+2i)$ and $(7+5i)$, and try multiplying them together using the rules of simple algebra, which say that you multiply the second brackets by 3, then by 2i and add the results together. So we get:

$$3(7 + 5i) + 2i(7 + 5i),$$
$$\text{which equals } (3 \times 7) + (3 \times 5i) + (2i \times 7) + (2i \times 5i)$$

If we don't do anything with the i s yet, but multiply and clump the terms together, we get

$$21 + 15i + 14i + 10i^2 = 21 + 29i + 10i^2$$

If we were trying to get a number that is part of the same system, another address on the same street map, we're not there yet, since we're looking for something of the form $a + ib$. We need one more step. If in the complex number $21 + 29i + 10i^2$ we substitute -1 for i^2, we get $21 + 29i + (10 \times -1)$, which equals $21 - 10 + 29i$, or $11 + 29i$. And $11 + 29i$ is another address on the map, 29 units along the 'street' that joins Camino Real at number 11.

A final point for now about these new numbers. They are not an alternative system to the real numbers, because the real numbers are included in them. Although every number is of the form $a + ib$, if $b = 0$ then we're left with a, the real part of the number, which is just a point on the line of real numbers.

In practice, these complex numbers are not always written out in full. Once the rules have been established for adding, subtracting, multiplying and dividing them, they can be manipulated more simply, and more abstractly, as single symbols. If we let z_1 stand for $a + ib$ and z_2 stand for $c + id$, for example, then we can write $z_1 + z_2$ instead of $(a + ib) + (c + id)$, or $z_1 z_2$ instead of $(a + ib) \times (c + id)$.

Let's just pause here to recap. The numbers we have all grown up with are just one of many number systems in mathematics. In fact, even within the familiar real numbers there are subsets such as the integers and the fractions (also called rationals, because they are depicted as ratios). But mathematicians have devised other number systems, which sometimes (but not always) obey the same rules. Complex numbers are at the heart of one of these systems, and we can sometimes manipulate them in similar ways to the real numbers.

The next step is crucial for understanding the Riemann zeta function. It is possible to use complex numbers in series such as the ones we looked at in Chapter 2. Although its meaning is not yet obvious, we can use z, as a complex number, in an expression such as $1/n^z$. Furthermore, we can put it in a series, as $\Sigma\, 1/n^z$. If we write this series out, remembering that $z = a + ib$ for some real numbers a and b, it looks like this:

$$1 + \frac{1}{2^z} + \frac{1}{3^z} + \frac{1}{4^z} + \cdots,$$

$$\text{which means } \quad 1 + \frac{1}{2^{(a+ib)}} + \frac{1}{3^{(a+ib)}} + \frac{1}{4^{(a+ib)}} + \cdots$$

There's no need to worry for now about understanding what this series actually means. That doesn't matter, since we've arrived at a key point in the story. This series is like the Euler zeta function, $\zeta(s)$, but with complex numbers instead of s. And *this* is the important change that Riemann made, to come up with the Riemann zeta function.

It's because Riemann used complex numbers instead of real numbers, with s being a number of the form $a + ib$, that the behaviour of the Riemann zeta function is far more mysterious than the behaviour of the Euler zeta function. And because the Riemann zeta function is crucial to replacing Gauss's guess for the number of primes less than any number n with an accurate answer, it has been the focus of probably more mathematical attention in the past hundred years than any other mathematical object.

And yet, not only is the Riemann zeta function entirely unfamiliar to 100 per cent of the educated population,* even the terms used to describe it are comprehensible only to a narrow, specialized group of people, largely mathematicians and physicists. This may not seem very surprising. After all, why *should* people remember maths once they have left school? Those who do remember are usually the ones who have understood and enjoyed it. Other people remember their history, geography, physics or English literature with a similar enthusiasm. The difference is that those subjects form part of the currency of everyday life in a civilized society, but maths, it seems, does not.

Things have not always been this way. Even making allowances for the more abstract nature of modern mathematics, there have been times in the past when mathematics was seen as a game that anyone could play

---

* 100 per cent doesn't always mean 'all' – see p. 92.

and enjoy. Somehow the symbolism was less of an obstacle. Here is a letter the poet Samuel Taylor Coleridge wrote to his brother, George:

Dear Brother,

I have often been surprised, that Mathematics, the quintessence of Truth, should have found admirers so few and so languid. – Frequent consideration and minute scrutiny have at length unravelled the cause – viz. – that though Reason is feasted, Imagination is starved; whilst Reason is luxuriating in it's [sic] proper Paradise, Imagination is wearily travelling on a dreary desart [sic]. To assist Reason by the stimulus of Imagination is the design of the following production. In the execution of it much may be objectionable. The verse (particularly in the introduction of the Ode) may be accused of unwarrantable liberties; but they are liberties equally homogeneal with the exactness of Mathematical disquisition, and the boldness of Pindaric daring. I have three strong champions to defend me against the attacks of Criticism: the Novelty, the Difficulty, and the Utility of the Work. I may justly plume myself, that I first have drawn the Nymph Mathesis from the visionary caves of Abstracted Idea, and caused her to unite with Harmony. The first-born of this Union I now present to you: with interested motives indeed – as I expect to receive in return the more valuable offspring of your Muse—

Thine ever,

S.T.C.

[Christ's Hospital,] *March 31, 1791.*[22]

Coleridge then offers in rhyming verse a proof of Euclid's First Proposition. This is a statement that using a ruler and compasses it is possible to construct a triangle with all three sides equal on any given base. It requires a certain amount of care in reading it as rhyme, and it begins:

This is now – this was erst,
Proposition the first – and Problem the first.

I
On a given finite Line
Which must no way incline;
To describe an equi-
-lateral Tri-
-A, N, G, L, E.

Now let A. B.
Be the given line
Which must no way incline;
The great Mathematician
Makes this Requisition,
That we describe an Equi-
-lateral Tri-
-angle on it:
Aid us, Reason – aid us, Wit! [23]

The proof goes on for several verses, not Coleridge's best, but that he has done it at all is surprising to most of us today.

There are two interesting things about Coleridge's poem. One is that it shows that it is actually possible to express mathematical ideas in words – but in a much more cumbersome way, which brings home the brevity that comes with symbols. The second is that a poet of the importance of Coleridge should interest himself in mathematics. Some of his friends were far less sympathetic. When Keats and Wordsworth, along with Charles Lamb, met for dinner in December 1817, twenty-six years after Coleridge wrote his youthful poem, they drank a famous toast to 'Newton's health and confusion to mathematics'.[24] Wordsworth had complained that Newton, though a romantic, was also 'a fellow who believed nothing unless it was as clear as the three sides of a triangle'[25] and that he had destroyed the beauty of the rainbow by explaining its origins in the optical properties of raindrops. Three years after the dinner and the toast, Keats, sharing this dyspeptic view, wrote:

Philosophy will clip an Angel's wings
Conquer all mysteries by rule and line
Empty the haunted air, and gnomed mine –
Unweave a rainbow...[26]

Coleridge's views and interests were more in keeping with the spirit of a time when educated people didn't scorn the heights of mathematics and even obtained some pleasure from its nursery slopes. Whenever Dr Johnson 'felt, or fancied he felt, his fancy disordered, his constant recurrence was to the study of arithmetic'.[27] Napoleon, 'when he had a few moments for diversion...not unfrequently [sic] employed them over a book of logarithms, in which he always found recreation'.[28] Thomas

Hobbes 'was 40 years old before he looked on geometry; which happened accidentally. He was in a gentleman's library, and Euclid's *Elements* lay open, at Pythagoras' Theorem. He read the proposition. "By God," sayd he, "this is impossible." So he reads the demonstration of it, which referred him back to such a proposition; which proposition he read. That referred him back to another, which he also read. At last he was demonstratively convinced of that truth. This made him in love with geometry.'[29] Even a humble nineteenth-century country parson 'sought his recreation in Lacroix's Differential Calculus and thus found intellectual refreshment for his calling'.[30]

Clearly, there was a time – now long gone – when educated men would see no shame in incorporating mathematics into their cultural lives. They had discovered that studying mathematics can be its own reward. Even today the occasional non-mathematician, well known in some other area of life, discovers for himself the pleasures of mathematics. Artie Shaw, a giant among jazz composers and performers, suddenly abandoned public performance and plunged into an intensive course of self-education. Among the subjects he explored was mathematics:

> All the time I was studying mathematics, I had an amazing feeling of certainty, a sense of logical absoluteness, so that, while I worked at the various branches of it, I seemed to be living in an atmosphere of complete and utter security. When it came to taking examinations, for instance, I can remember the absolute confidence with which I approached them. There was never any doubt in my mind as to the results. And the results themselves bore me out. There was never any question of passing or not passing – for you either knew the subject (in which case there was no doubt about it at all and you got a perfect score of 100%), or you didn't (in which case you had no business taking the exam until you had finished learning what you had to in order to know it); and because these mathematical studies were giving me, in some curious way, the only sense of actual security (on any level) I had ever known, I was about as happy, intellectually, as I had ever been in my entire life up to that point.[31]

It's a rare for non-mathematicians to discover for themselves the intellectual satisfaction that mathematics can bring. But for practising mathematicians, that is the earliest and most enduring motivation for pursuing their profession.

4 Indian summer

> The last thing I want you to do is throw up your hands and exclaim 'Here is something unintelligible, some mysterious manifestation of the immemorial wisdom of the East!'...The picture I want to present to you is that of a man with whom one could drink tea...a rational human being who happened to be a great mathematician.
>
> G.H. Hardy, on Ramanujan [32]

In 1900, the Riemann Hypothesis was given new prominence in a famous lecture by the great German mathematician David Hilbert. As E.T. Bell, an eloquent but cranky historian of mathematics, pointed out with reference to Hilbert, mathematicians, however great, never become as well known as other scientists, such as Einstein:

> It is a fair guess that out of 100,000 persons picked at random on the streets of New York, or Chicago, or London, or Paris, or Moscow, or Tokyo, not one would know the name of the man who professional mathematicians almost unanimously considered was the foremost member of their guild since about 1912. He died in 1943, inactive; but his fame is secure even if the average man in the street (or in a cultured society) is never likely to hear of him. Of the random hundred thousand, many would instantly name a theoretical physicist who deeply resents being called a mathematician. [33]

Hilbert had been invited to give a keynote address at the second International Congress of Mathematicians, held in Paris in the summer of 1900. For such a momentous occasion, the turnout was disappointing. A thousand mathematicians from round the world, many accompanied by family members, had turned up in Paris – partly, it has to be said, for the Centennial Exhibition being held there – but only two hundred and fifty or so made it into the lecture hall on a swelteringly hot morning in August. The figures may have been lower than expected because Hilbert had procrastinated and submitted the text of his lecture too late for it to be included as the opening address, as had originally been intended. Given the significance this talk was to assume over the next hundred years, there must be many mathematicians who later wished they had been there.

Constance Reid, Hilbert's biographer, described Hilbert's appearance that day:

> The man who came to the rostrum that morning was not quite forty, of middle height and build, wiry, quick, with a noticeably high forehead, bald except for wisps of still reddish hair. Glasses were set firmly on a strong nose. There was a small beard, a still somewhat straggly moustache and under it a mouth surprisingly wide and generous for the delicate chin. Bright blue eyes looked innocently but firmly out from behind shining lenses. In spite of the generally unpretentious appearance of the speaker, there was about him a striking quality of intensity and intelligence... Slowly and carefully for those who did not understand German well, he began to speak.[34]

But what was he about to say that was of such importance?

Hilbert decided to use the occasion of the new century to look at the topic of unsolved problems in mathematics, and to explore the value to mathematics of problems in general. He believed that 'A branch of science is full of life as long as it offers an abundance of problems; a lack of problems is a sign of death.'

> It is difficult and often impossible to judge the value of a problem correctly in advance [Hilbert said, as he began to develop his ideas to the audience] for the final award depends upon the gain which science obtains from the problem. Nevertheless, we can ask whether there are general criteria which mark a good mathematical problem. An old French mathematician said: 'A mathematical theory is not to be considered complete until you have made it so clear that you can explain it to the first man whom you meet on the street.' This clarity and ease of comprehension, here insisted on for a mathematical theory, I should still more demand for a mathematical problem if it is to be perfect; for what is clear and easily comprehended attracts, the complicated repels us. Moreover, a mathematical problem should be difficult in order to entice us, yet not completely inaccessible, lest it mock our efforts. It should be to us a

guidepost on the tortuous paths to hidden truths, ultimately rewarding us by the pleasure in the successful solution.[35]

If there were any 'men (or women) on the street' sitting in the hall, it is unlikely that they would have found much 'clarity and ease of comprehension' in the list of ten unsolved problems Hilbert described, problems he hoped would be solved some time during the twentieth century. (He had actually chosen twenty-three, which were included in the text of his talk, but he only had time to speak about ten of them.) These were not mere brain-teasers from the puzzle pages of newspapers. Each of them came from some key field of mathematics at the time. If they were to be solved, their solutions, Hilbert believed, would advance that field in new and promising directions. Here are the first few items on Hilbert's list:

1. Cantor's problem of the cardinal number of the continuum.
2. The compatibility of the arithmetical axioms.
3. The equality of the volumes of two tetrahedra of equal bases and equal altitudes.
4. The problem of the straight line as the shortest distance between two points.

Even though problem 4 is expressed in words which, taken on their own, we might understand, it was clearly more important than it sounds. Why should there be any problem about the straight line as the shortest distance between two points? In fact, when it was rephrased by Hilbert in his published version of the talk it lost all clarity and became 'Find geometries whose axioms are closest to those of Euclidean geometry if the ordering and incidence axioms are retained, the congruence axioms weakened, and the equivalent of the parallel postulate omitted', which is perhaps nearer to the inscrutability non-mathematicians expect from the frontiers of mathematics.

> The problems mentioned [Hilbert told his audience] are merely samples of problems; yet they are sufficient to show how rich, how manifold and how extensive the mathematical science is today; and the question is urged upon us whether mathematics is doomed to the fate of those other sciences that have split up into separate branches, whose representatives scarcely understand one another and whose connection becomes ever more loose. I do not believe this nor wish it... The organic unity of

mathematics is inherent in the nature of this science, for mathematics is the foundation of all exact knowledge of natural phenomena. That it may completely fulfil this high destiny, may the new century bring it gifted prophets and many zealous and enthusiastic disciples![36]

We can place Hilbert's problems into three categories: (1) problems that were solved in Hilbert's lifetime – he died in 1943; (2) problems that were eventually solved, with a lot of hard work and sometimes nearly a hundred years after Hilbert's talk; and (3) problems that have still not been solved, and may never be.

Into the first category falls Hilbert's problem 3, which was solved within two years of his lecture by one of his own students. Problem 4 fell in 1910, problem 11 in 1923, problem 17 in 1926, problem 9 in 1927, problem 2 in 1931; and problem 7 in 1934.

The second category of Hilbert's problems, those solved after Hilbert's death and after a lot of hard work, includes the Tenth Problem, about what are called Diophantine equations. These are equations whose unknowns are whole numbers. The American mathematician Julia Robinson devoted years of work to Hilbert's Tenth Problem and left an affecting account of its importance in her life:

> Throughout the 1960s, while publishing a few papers on other things, I kept working on the Tenth Problem, but I was getting rather discouraged. It was the custom in our family to have a get-together for each family member's birthday. When it came time for me to blow out the candles on my cake, I always wished, year after year, that the Tenth Problem would be solved – not that I would solve it, but just that it *would* be solved. I felt that I couldn't bear to die without knowing the answer.
>
> On January 4, 1970, two months before his twenty-third birthday, Yuri Matijasevich was finally able to prove that… the long awaited solution of Hilbert's tenth problem turned out to be negative. [Matijasevitch, a Russian mathematician working in Leningrad, had used a conjecture of Robinson's in his proof.] I was so excited by the news, that I wanted to call Leningrad right away to find out if it were really true… Just one week after I had first heard the news, I was able to write to Matijasevich: '…now I know it is true, it is beautiful, it is wonderful. If you really are 22, I am especially pleased to think that when I first made the conjecture you were a baby and I just had to wait for you to grow up!' That year when I went to blow out the candles on my cake, I stopped in mid-breath, suddenly realizing that the wish I had made for so many years had actually come true.[37]

Also solved after Hilbert's death was Fermat's Last Theorem, proved to be correct as recently as 1996, by Andrew Wiles. This problem is nearest in complexity and difficulty to the Riemann Hypothesis and, like the Riemann Hypothesis, has been the focus of intense and single-minded efforts by number theorists over the years. Wiles first came across the problem when he was ten years old, and after he'd became a mathematician he was determined to find a proof single-handed. He refused to collaborate with anyone during three decades of thinking about the problem, seeking the satisfaction – and perhaps the fame – that would come from doing it all on his own.

The third category of Hilbert's problems, the ones that have still not been solved, included Hilbert's problem 8, stated simply as 'Problems of prime numbers (including the Riemann hypothesis).'

Hugh Montgomery describes what lay behind Hilbert's bald description. 'Hilbert had one blanket problem that asked all possible questions about primes. It would be very difficult to say you'd solved that unless you'd proved the Riemann Hypothesis and Twin Primes conjecture and so on, but…I think at that time one would have expected that it would fall sooner rather than later. There was no reason to think it was going to be one of these resistant problems that sticks around. It's very difficult to speculate about the future unfolding of mathematics – you look at some problems that are unsolved at one time and can make guesses about what order things will follow based on recent progress and ideas in the field and apparent difficulties and so on, but such speculations are usually very wide of the mark. Hilbert gave a famous address in 1931 to the German Mathematical Society, and he said roughly that he thought the Riemann Hypothesis would be proved in his lifetime, but he thought it would be much longer – I don't know if he said people would be visiting the stars – but much longer before one knew that $2^{\sqrt{2}}$ would be transcendental.* 2 to the square root of 2 was proved to be transcendental three years later by two different people by independent methods. Hilbert died in 1943,

---

* A transcendental number is one that is neither an integer, nor a fraction, nor the root of an algebraic equation such as $2 - x^2 = 0$. There are several very well-known transcendental numbers, like π and e, the base of natural logarithms, but in fact there are many more lesser-known ones. It's possible to prove that there are far more transcendental numbers than any other numbers on the line of real numbers. The number $2^{\sqrt{2}}$ (2 raised to the power of the square root of 2) is therefore an integer raised to the power of an irrational number.

and here we are now and we still don't have the Riemann Hypothesis.'

For Hilbert, the Riemann Hypothesis became the most important of all his problems, if we are to believe a story often told in mathematical circles. According to German legend, after the death of Barbarossa, the Emperor Frederick I, during a Crusade, he was buried in a faraway grave. It was rumoured that he was still alive but asleep, and would wake one day to save Germany from disaster, even after five hundred years. Hilbert was once asked, 'If you were to revive, like Barbarossa, after five hundred years, what would you do?' He replied, 'I would ask, "Has somebody proved the Riemann Hypothesis?" ' [38]

When Hilbert gave his keynote lecture in 1900, three mathematicians whose lives and work were to become famously intertwined were twenty-three, fifteen and thirteen years old. They were, respectively, G.H. Hardy, Jack Littlewood and Srinivasa Ramanujan.* Hardy first saw Ramanujan's mathematics in a neat, handwritten letter he received from Madras on 16 January 1913. Hardy, as a leading English pure mathematician, was used to getting letters from cranks containing proofs of such unprovable mathematical conundrums as squaring the circle and trisecting an angle. The maths in Ramanujan's letter was of a higher order than that, but Hardy can still be forgiven for not having dropped everything to analyse it in detail, since one of Ramanujan's offerings was 'a function which *exactly* represents the number of prime numbers less than x' – a function which, if genuine, would have been equivalent to the Prime Number Theorem, then only recently proved.[39] Furthermore, the proof Ramanujan gave for his statement was – if correct – entirely different and in some ways much simpler. The likelihood of such mathematical brilliance turning up out of the blue from a young clerk in the port of Madras was pretty low. As it turned out, Ramanujan's proof of the Prime Number Theorem *was* wrong, but much of the rest of the maths in his letter was brilliant. Eventually, Hardy arranged for Ramanujan to come to Cambridge where, from 1914 to 1919, the two men, along with Jack Littlewood, Hardy's closest collaborator, worked on the theory of numbers, which is fundamental to understanding the Riemann Hypothesis.

Both Hardy and Littlewood were passionate about maths, almost to the exclusion of everything else, apart from cricket in Hardy's case.

---

* Pronounced Ra-*man*-ujan.

They were both Fellows of Trinity College, Cambridge. Founded in 1546, the college was a haven where often unworldly academics could work away at their specialist subjects untroubled by the need to do the shopping, make their beds or walk more than a few yards to and from a dining hall where good food and drink were supplied. They didn't come much more unworldly than Hardy. According to C.P. Snow, then a Fellow in English at Christ's College, Hardy 'had a morbid suspicion of mechanical gadgets (he never used a watch), in particular of the telephone. In his rooms in Trinity or his flat in St George's Square, he used to say, in a disapproving and slightly sinister tone, "If you *fancy yourself* at the telephone, there is one in the next room." '[40]

Even in the 1980s, when I and a colleague were visiting Trinity College in search of information about Ramanujan for a television programme, one of the dons said to my colleague, 'Television? Television? I believe we have one of those in the corner of the Senior Common Room, but it's not often *lit*.' Such an ability to confuse a television set with a stove was worthy of Hardy himself.

Hardy and Littlewood wrote a number of key papers in number theory, some together, some separately. Harald Bohr, a Danish mathematician and close friend of both of them, remarked:

> I may report what an excellent colleague once jokingly said: 'Nowadays, there are only three really great English mathematicians: Hardy, Littlewood and Hardy–Littlewood.' The last refers to the marvellous collaboration through the years between these two equally outstanding scientists with their very different personalities. This co-operation was to lead to such great results and to the creation of entirely new methods, not least in the theory of numbers, that to the uninitiated, they almost seemed to have fused into one.[41]

Although they both lived in the same college, they often communicated by letter, delivered by college porters, in which they addressed each other as 'Dear Hardy' and 'Dear Littlewood'. They were extremely serious about mathematics but could show bursts of donnish wit, as in this comment by Littlewood on Hardy:

> The letters L, R, (in a Dedekind section) for which a generation of students is rightly grateful, were introduced by me. In the first edition of *Pure*

*Mathematics** they are T, U. The latest editions have handsome references to me, but when I told Hardy he should acknowledge this contribution (which he had forgotten) he refused on the ground that it would be insulting to mention anything so minor. (The familiar response of the oppressor: what the victim wants is not in his own best interests.)[42]

But there was no frivolity when it came to mathematics. These were people for whom numbers and what could be done with them were the most exciting and important things in the universe. Describing his student days Littlewood wrote: 'A few weeks before the examination, in the Easter term, I first came across the early volumes of the Borel series [a series of maths textbooks], and it was these, in cold fact, that first gave me an authentic thrill: series of positive terms, divergent series, and the volume on integral functions.'[43] This passion was accompanied by a scrupulous honesty, not just in the execution of mathematics, as shown by Littlewood's comment about a problem he had been trying to solve during a university maths test:

> It did not come out, nor did it on a later attack. I had occasion to fetch more paper; when passing a desk my eye lit on a heavy mark against the question. The candidate was not one of the leading people, and I half unconsciously inferred that I was making unnecessarily heavy weather; the question then came out fairly easily. The perfectly high-minded man would no doubt have abstained from further attack; I wish I had done so, but the offence does not lie very heavily on my conscience.[44]

Littlewood's first contact with the Riemann Hypothesis came within four years of Hilbert's lecture. At a time when Littlewood was still an undergraduate and unaware of the significance – or the difficulty – of the problem, he so impressed his tutor with a fifty-page paper on 'Entire functions of order zero' that he was next asked to prove the Riemann Hypothesis. Littlewood later described his reactions to being given this task:

> As a matter of fact this heroic suggestion was not without result. I had met the zeta function in Lindelöf[†] but there is nothing there about primes, nor had I the faintest idea there was any connexion; for me the Riemann

---

* A book written by Hardy.
† A mathematical text.

Hypothesis was famous, but only as a problem in integral functions; and all this took place in the Long Vacation when I had no access to literature, had I suspected there was any.

I remembered the Euler formula;* it was introduced to us at school, as a joke (rightly enough and in excellent taste)... In the light of Euler's formula it is natural to study $P(s) = \Sigma\, p^{-s}$. I soon saw that if the Prime Number Theorem were true 'with error about $\sqrt{x}$' the Riemann Hypothesis would follow. Now at that time, and for anyone unacquainted with the literature, there was no reason to expect any devilment in the primes... So I started off in great excitement and confidence, and only after a week or two of agony came to realize the true state of things.[45]

Littlewood became 'infatuated with the problem' and spent much of his professional life working on the Riemann zeta function and the Riemann Hypothesis, proving several related results in the process. Sixty-five years after his first undergraduate contact with the problem, Littlewood sought the help of Bela Bollobas, a Hungarian fellow of Trinity, in dealing with one of many 'proofs' which found their way onto his desk for consideration:

In 1971 a paper was submitted to the London Mathematical Society, claiming a proof of the Riemann Hypothesis. The author sent a copy to Littlewood as well. For a while Littlewood was patient. But months went by and the editors still held on to the paper. Was it perhaps correct? Littlewood became more and more demanding so I hawked the paper around the mathematics department and even wrote to the editors of the [Society's] Journal. I had no luck: nobody was willing to waste his time trying to penetrate the complicated (and somewhat old fashioned) formulae. Eventually I had to give in to Littlewood's demand and we began to read the paper together. I knew nothing about the formulae and I was amazed how well *he* knew them. After a few hours of painstaking work he was relieved to find a mistake. But that isn't the end of the story. A week or so later I got the referee's report, pointing out a mistake which was *not* there.[46]

For Hardy too, the Riemann Hypothesis was a crucial but elusive goal. One year, at the top of Hardy's list of New Year's resolutions was 'Prove

---

* The formula linking the whole numbers and the primes, described in Chapter 2.

the Riemann Hypothesis'. There were five more, whose unlikelihood made the Riemann Hypothesis proof sound easy in comparison. They were:

2. Make 211 not out in the fourth innings of the last test match at the Oval. (211 has no particular significance in cricket, but it *is* a prime number.)

3. Find an argument for the nonexistence of God which shall convince the general public.

4. Be the first man on the top of Everest.

5. Be proclaimed first president of the U.S.S.R. of Great Britain and Germany.

6. Murder Mussolini.[47]

When this witty, ascetic intellectual Englishman met Srinivasa Ramanujan, his life changed.

When I lived in London, I often drove along the Upper Richmond Road, and I always thought of Hardy and Ramanujan as I passed the end of a turning called Colinette Road. It's an ordinary suburban street, with family houses, many now divided into flats. But Number 2 Colinette Road may have been an ordinary address, but the house itself was the scene of a piece of mathematical history. In 1919 it was a convalescent home, a place where people who had been seriously ill and were now on the mend could recover in peace and tranquillity. Ramanujan was visited here by Hardy one day. Ramanujan had become weak and debilitated after several years of an indefinable malaise, perhaps tuberculosis, which some said was exacerbated by an inability to find the ingredients for his Hindu diet in wartime Britain.

It was said of Ramanujan that 'each of the positive integers was one of his personal friends',[48] and when Hardy turned up to visit him, Ramanujan showed that this was quite true. Hardy had come in a London taxi, number 1,729, and as he went into the room where Ramanujan was lying, he remarked that the number of the taxi was rather a dull one. To which Ramanujan replied, 'No, it is a very interesting number. It is the smallest number expressible as the sum of two cubes in two different ways.'

This story, now part of the folklore of mathematics, was told by

Hardy in his collected works, and repeated by C.P. Snow in a foreword to *A Mathematician's Apology*, a short popular book that Hardy wrote about his life and his maths.[49] It's a story that in recent years has acquired an almost mystical aura.

'I'd always been very fascinated with Ramanujan and with this story about the number 1,729,' wrote mathematician David Ash, who has 'studied with several Tantric Buddhist masters, including Rama/ Dr. Frederick Lenz'[50] and is a recent contributor to the folklore.

'In late 1992,' Ash continued, 'I dropped my mentor, who was going away on a short vacation, off at the train station in San Jose, California. We chatted for a bit at the station, and then I returned to my car. I was very interested to notice a taxicab outside the station with a license plate which consisted of a few letters and then the numbers 1,729. It was the first time I'd ever seen a taxi with the numbers 1,729, and given my fascination with Ramanujan I thought it was a powerful omen.'[51]

When I first heard the 1,729 story, I – like many people – marvelled at the quickness of Ramanujan's mind, imagining it to be like a very fast fruit-machine, zipping through the process of adding all pairs of cubes, until 'ching!' – there were the two pairs of numbers that produced the result, $1^3 + 12^3$ and $10^3 + 9^3$. What's more, there was no smaller number with which you could do the same thing. In fact, as I read more about this number, the biggest puzzle was not how Ramanujan managed to recognize its unusual nature, but why Hardy, one of England's greatest mathematicians, hadn't.

The number cropped up in a thread of correspondence about maths on the Edge Foundation website. Charles Simonyi, a leading computer engineer at Microsoft, wrote:

As to the Ramanujan number 1,729: has anybody noticed the same number appearing in [Richard] Feynman's book [*'Surely You're Joking, Dr Feynman?'*] when he competes using mental arithmetic with the Japanese abacus master? After losing on simple arithmetic the problem given was to calculate the cube root of 1,729! Feynman apparently had not heard the Ramanujan story because he does not mention it at all in the book: his insight was to remember from his wartime engineering days that one cubic foot equals 1,728 cubic inches! So as the Japanese sweated on the abacus he slowly emitted the obvious digits, 1, 2, 0, while feverishly trying to approximate a little bit of the rest using a power series. Also, the octal* representation of this number 3,301 was for a long time the secret password to the central computer of Xerox Parc. Maybe this will spur someone on to more insights.[52]

This was followed by a comment from Stanislas Dehaene, a neuro-psychologist:

> The story about Richard Feynman remembering that 1,728 is 12 × 12 × 12 is particularly interesting, because it strengthens the idea that, indeed, 1,728 is a very recognizable number, and that many people – including at least Ramanujan, Feynman, Simonyi, myself, and surely scores of others – have it stored in their mental number lexicon. This does not in the least diminish my admiration for Ramanujan's feats, but it does make him seem more human and understandable.[53]

So, someone familiar with cubes would reason as follows. You'll recognize straight away that 729 is a cube (9^3). Also, 1,000 is 10^3, so 1,729 is already interesting as the sum of two cubes. And when 1,729 is only one unit away from 1,728, another known cube, you can easily see that 1,729 is the sum $12^3 + 1^3$. The only truly mathematical piece of the equation, so to speak (since the rest is based on rote memory of cubes), is knowing – as Ramanujan did – that 1,729 is the *smallest* number expressible as the sum of two cubes in two different ways. That *was* quite clever.

For a mind like Ramanujan's, the unexpected complexity of the whole numbers – the games that can be played with them, the questions that can be asked about them, the speed with which clarity turns into difficulty – is powerfully seductive. This is true of many mathematicians, if to a lesser degree. In many cases it's what got them started on maths in the first place.

The Hardy–Littlewood mathematical papers are incomprehensible to the non-mathematician, but occasionally they give an insight into the peaks of excitement – the 'authentic thrill' to be had from pursuing the solution to a mathematical problem. One account of Littlewood's, featuring Hardy and Ramanujan and written in lively and emotional language, is worth looking at closely even though it is littered with formulae and equations that contain symbols which have deeply specialized meanings. At the heart of the account is a piece of higher

---

* Using 8 as a base instead of 10, as in the decimal system. So $1729 = (3 \times 8^3) + (3 \times 8^2) + (0 \times 8) + 1$.

mathematics, but it actually deals with a concept that emerges from the simplest possible mathematical procedure, the addition of whole numbers. Just as all the complexity of the Riemann Hypothesis arises from a simple question about prime numbers, one of Ramanujan's contributions to higher mathematics comes from asking an even simpler question to do with 'sums' of the sort that any child can do. Littlewood's comments on how Ramanujan arrived at his insight are worth exploring in a little detail.

We need a little specialist terminology to start with. Ramanujan was interested in what are called 'partitions'. This is one of the easiest concepts in number theory to describe, but it led Ramanujan and Hardy to produce a formula which, to the uninitiated, is of fearsome complexity. But, then, that happens in mathematics all the time. As we will see with the Riemann Hypothesis, one of the things that makes number theory so captivating to mathematicians is the hidden depths that lurk beneath a placid and sometimes obvious surface.

A partition of a whole number N is the number of different ways you can add other whole numbers together to make N. For example, you can get the result 6 by adding 3 and 3; 2 and 4; 5 and 1; 1,2 and 3; 1, 1, 1, 1, 1 and 1; 1, 1, 1 ,1 and 2; 1, 1, 1 and 3; 1, 1 and 4; 2, 2, and 2; and 2, 2, 1 and 1; or by taking 6 on its own. So there are eleven different ways to get the result 6 by adding together smaller integers. To get the result 5, there are only seven ways – by adding 1, 2, 3 and 4 in different combinations. But with the number 20 there are 627 ways. Each of the sums that add up to 6 is called a 'partition' of 6. Because there are eleven partitions of 6, mathematicians write $p(6) = 11$. Similarly, $p(5) = 7$ and $p(20) = 627$.

An interesting fact about the partitions of a number is that some of them come in pairs which seem, at a glance, to have no connection with one another, but on closer inspection are found to be intimately related.

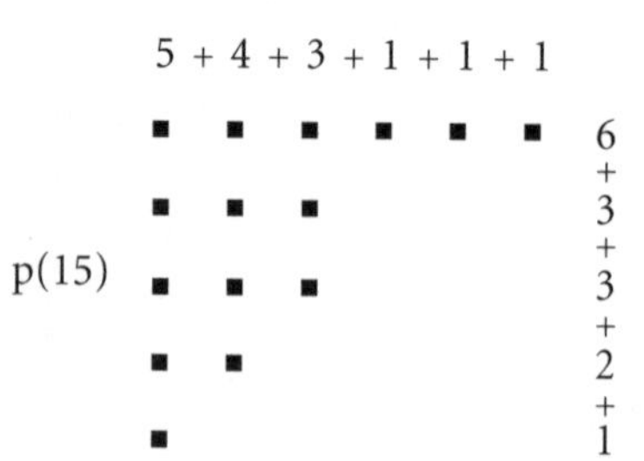

FIGURE 4 How simple sums can lead to complicated mathematics. A partition of a number is how many different ways you can add smaller numbers together to make that number. Here are two partitions of 15, related to each other through the pattern of dots.

One partition of 15 is $6+3+3+2+1$; another is $5+4+3+1+1$. But if you write the first partition out as a series of dots, as in Figure 4, you'll see that if you read downwards instead of across you'll get the second partition. You can even make a deduction about the relationship between the two. It is clear that the largest part in either of the partitions is equal to the number of rows in the other.

So far, then, there's nothing too mathematical to get confused about, even if there doesn't seem too much to get excited about either. The table on the right gives the partitions of the first thirty integers. When Ramanujan needed to know the value of $p(n)$ for a particular number n, he turned to a table compiled by another Cambridge resident, one Major Percy MacMahon, formerly of the Royal Artillery regiment. Major MacMahon liked nothing better than to calculate how many different ways you could add integers to get, say, $p(100)$. Fortunately there was a method that avoided working out and writing down $190,569,292$ sums – the correct value – though it was almost as cumbersome.

When a mathematician is confronted with a series of terms like this list of partitions, the first question that immediately comes to mind is whether there is a formula that will generate these numbers. In other words, does $p(n) = n^2$, say (obviously not). How about $p(n) = n$? That seems to work for $n = 1$, 2 or 3, but it starts to veer off above 3. Not really much use either.

Hardy and Ramanujan asked themselves this question, worked away at it, and came up with a formula that was amazingly accurate. If you substitute any whole number in their formula and work it out, the answer will be $p(n)$. When Littlewood came to review the book of collected papers by Ramanujan, he singled out the paper on $p(n)$ as an example of the supreme quality of Ramanujan's mathematics. The paper showed, as examples, how the formula Ramanujan had arrived at with the help of Hardy calculated

p(1)	1
p(2)	2
p(3)	3
p(4)	5
p(5)	7
p(6)	11
p(7)	15
p(8)	22
p(9)	30
p(10)	42
p(11)	56
p(12)	77
p(13)	101
p(14)	135
p(15)	176
p(16)	231
p(17)	297
p(18)	385
p(19)	490
p(20)	627
p(21)	792
p(22)	1,002
p(23)	1,255
p(24)	1,575
p(25)	1,958
p(26)	2,436
p(27)	3,010
p(28)	3,718
p(29)	4,565
p(30)	5,604

p(100) and p(200) accurately, giving the correct values as 190,569,292 and 3,972,999,029,388. Littlewood's comments reveal the kind of excitement and admiration that others might express on hearing a virtuoso pianist, looking at a masterpiece in an art gallery, or watching a consummate athlete (my italics):

> The reader does not need to be told that this is *a very astonishing theorem*, and he will really believe that the methods by which it was established involve *a new and important principle*, which has been found very fruitful in other fields. The story of the theorem is *a romantic one*. [Littlewood begins to describe the process] From this point *the real attack begins*...But from now to the very end Ramanujan always insisted that much more was true than had been established: 'There must be a formula with error $O(1)$'. This was his most important contribution; it was both *absolutely essential and most extraordinary*. A severe numerical test was now made which elicited *the astonishing facts* about p(100) and p(200). Then v was made a function of n; this was *a very great step*, and involved new and deep function-theory methods that Ramanujan obviously could not have discovered by himself. The complete theorem thus emerged. But the solution of the final difficulty was probably impossible without one more contribution from Ramanujan, this time a perfectly characteristic one. As if its analytical difficulties were not enough, the theorem was *entrenched also behind almost impregnable defences* of a purely formal kind. The form of the function $\psi_q(n)$ is a kind of indivisible unit; among many asymptotically equivalent forms it is essential to select exactly the right one. Unless this is done at the outset, and the $-\frac{1}{24}$ (to say nothing of the d/dn) is *an extraordinary stroke of formal genius*, the complete result can never come into the picture at all. There is, indeed, a touch of real mystery. If only we knew there was a formula with error $O(1)$, we might be forced, by slow stages, to the correct form of ψ_q. But *why was Ramanujan so certain there was one?* Theoretical insight, to be the explanation, had to be of an order *hardly to be credited*. Yet it is hard to see what numerical instances could have been available to suggest so strong a result...there seems no escape, at least, from the conclusion that the discovery of the correct form has *a single stroke of insight*. We owe the theorem to *a singularly happy collaboration of two men*, of quite unlike gifts, in which each contributed the best, most characteristic, and most fortunate work that was in him. Ramanujan's genius did have this one opportunity worthy of it.[54]

Without understanding any of the mathematical terms in this paragraph, we can share at least some of Littlewood's excitement at seeing Ramanujan in action. But if we look in more detail at the substance of what Littlewood wrote, we can see something else which I think lies at the root of the fascination of pure mathematics. Remember how the Hardy–Ramanujan episode started – with an attempt to find out how many different ways you could add numbers to produce another number. A schoolchild could work out the value of $p(n)$ for $n = 1, 2, 3, 4, 5$, until he or she got bored or made a trivial mistake. But the formula that provides the general answer without working it out for each number – the subject of the Hardy–Ramanujan paper – was amazingly complex. Here it is:

$$\frac{1}{2\sqrt{2}} \sum_{q=1}^{v} \sqrt{q}\, A_q(n)\psi_q(n),$$

where

$$A_q(n) = \sum \omega_{p,q} \exp\left(-2np\pi\mathrm{i}/q\right)$$

the sum being over ps that are prime to q and less than q, $\omega_{p,q}$ is a certain $24q$-th root of unity, v is of the order of $\sqrt{n}$, and

$$\psi_q(n) = \frac{\mathrm{d}}{\mathrm{d}n}\left(\exp\left\{C\sqrt{\left(n - \tfrac{1}{24}\right)}\big/q\right\}\right), \quad C = \pi\sqrt{\tfrac{2}{3}}$$

'The reader does not need to be told that this is a very astonishing theorem', wrote Littlewood. Not the least astonishing thing about such a complex expression is that if you substitute, say, 6 for n you get the answer 11 (which is $p(6)$, as we saw earlier). It's almost like magic – an incantation built from a number of higher mathematical expressions yet which transforms one simple whole number into another.

There's little that's familiar in this whopper of a formula. Apart from a couple of square root signs, it is a mishmash of Greek letters – v, Σ, ω, ψ and π – the expression $\mathrm{d}/\mathrm{d}n$, which is from calculus, and the letter e, which – as we have seen – stands for a unique number which crops up a lot in number theory. But it's worse than that. We can accept that these symbols stand for numbers and processes that we don't use ourselves, so there's no need to know them. But if you look more carefully you'll find expressions that sound meaningless or nonsensical. What, for example, is 'a certain 24q-th root of unity'? Taken literally, it means a number which, multiplied together 24q times would equal 1. But *any* root of 1 is surely just 1, isn't it? The square root of 1 is one, because $1 \times 1 = 1$. The cube root of one is one ($1 \times 1 \times 1 = 1$), and in fact the 48th root of one is also 1, because $1^{48} = 1$.

Another element of this formidable formula which can make the non-mathematician go pi-eyed is embedded in the expression $e^{2np\pi i/q}$, or, in larger print which is easier to read,

$$e^{2np\pi i/q}$$

The letter e is a mathematical constant, something like π in the sense that it stands for a fixed number which crops up in many different fields of mathematics and has a numerical value of about 2.718.

The letter n in this expression stands for the number we're trying to find $p(n)$ for. We can leave p aside for the moment, because it's not the same thing as the p in $p(n)$ and it would all get too confusing. π is another fixed number, the ratio of the circumference of a circle to its diameter; and then we have the letter i, which as we have seen has the surprisingly useful function of representing the square root of minus one.

There is a sure-footedness about the steps that led to this formula which was so far away in appearance from the simple partitions that posed the initial question, 'What is the rule for $p(n)$?' But then the three men were totally at ease with number theory. They give the impression of effortless superiority when you read their finished works, especially for those of us for whom any mathematical expression more complicated than $9 \times 7 = 63$ can induce head-scratching and even revulsion. But even things that look difficult aren't necessarily so.

I came across a problem that looked to me as difficult as the Riemann Hypothesis itself, and yet it was set for schoolboys. As Hardy analysed it, he demonstrated what was clearly an intuitive grasp of the nuances of numbers, powers and margins of error. Surprisingly often in mathematics, the art is in knowing when and how to abandon the search for extreme accuracy and make approximations that allow the answer to come with much less effort. The problem, posed in Hardy's *Collected Papers*, is this:

Find an approximation to the large positive root of the equation

$$e^{e^x} = 10^{10} x^{10} e^{10^{10} x^{10}}$$

In other words, what number substituted for x will make the left side equal the right?

Hardy describes the method of solving this problem. 'The points to observe are (i) that the factor $10^{10} x^{10}$ proves to be of no importance whatsoever, and (ii) that it is futile to try to be very accurate in the early stages of the work...The great weakness of boys confronted with a numerical problem is that they cannot see where accuracy is essential and where it is entirely useless.' By the way, the answer to the problem is that x is somewhere between 63 and 67, and, says Hardy, 'a closer approximation could be found with a little trouble'.[55]

The Hardy–Ramanujan–Littlewood collaboration is a well-known episode in the recent history of mathematics. For a few years at the beginning of the twentieth century, three mathematical brains of unusual talent came together to spin off theorem after theorem, function after function, all for the sheer pleasure and excitement of doing it. The mathematics Ramanujan had produced on his own in India and the new results he produced in collaboration with Hardy were to provide food for thought for mathematicians long after Ramanujan's premature death at the age of thirty-three in 1920. For Hardy and Littlewood, in the years after Ramanujan's death, the Riemann Hypothesis was always in the background and sometimes in the foreground.

George Pólya, a talented Hungarian mathematician and friend of Hardy, described his friend's passion for the Riemann Hypothesis, which showed itself during Hardy's regular visits to his Danish colleague, Harald Bohr. (It helps to know that Hardy was a committed atheist.)

They had a set routine. First they sat down and talked, and then they went for a walk. As they sat down, they made up and wrote down an agenda. The first point of the agenda was always the same: 'Prove the Riemann hypothesis.'...Hardy stayed in Denmark with Bohr until the very end of the summer vacation, and when he was obliged to return to England to start his lectures there was only a very small boat available. The North Sea can be pretty rough and the probability that such a small boat would sink was not exactly zero. Still, Hardy took the boat, but sent a postcard to Bohr: 'I proved the Riemann hypothesis, G.H. Hardy.' If the boat sinks and Hardy drowns, everybody must believe that he has proved the Riemann hypothesis. Yet God would not let Hardy have such a great honour and so he will not let the boat sink.[56]

But in spite of the rosy glow of their achievements, all of Hardy and Littlewood's attempts to prove the Riemann Hypothesis failed, and the baton was to pass to a new generation of mathematicians.

'Very probably'

Where are the zeros of zeta of *s*?
G.F.B. Riemann has made a good guess,
They're all on the critical line, said he,
And their density's one over $2\pi \log t$.

Tom Apostol

The Riemann Hypothesis inspires a peculiar passion in the mathematicians who work on it. It's a passion that was often sparked almost as soon as they had grasped what number theory is all about. Henryk Iwaniec is an American mathematician of Polish origin, now working at Rutgers University in New Jersey. He is short and intense, fast-talking, with a heavy accent. When I first met him he explained to me that I might not find him very interesting because he 'was not colourful'. But, like most enthusiasts, when talking about the subject he loves he is unstoppable and, occasionally, even colourful. His life has been spent studying prime numbers. In 1997 he and John Friedlander, of the University of Toronto, devised a mathematical proof which was greeted with astonishment by other mathematicians. They work with what are called sieve methods, which divide the whole numbers into two groups, one of which contains a higher proportion of primes than average and the other a lower proportion – or even none at all.

To see how it works, imagine a very simple sieve for numbers which are squares. If you retained all whole numbers ending in 1, 4, 5, 6 and 9 and discarded the rest, this would be a sieve for the squares. You would know two things after this process: first, that there were no squares in the discarded numbers; and, second, that there would be a higher proportion of squares among the remaining numbers than there are in *all* the whole numbers. In that sense your 'sieve' will have sieved out the non-squares and concentrated the squares in the remaining numbers. All of this is based on the fact that squares must end with one of the numbers

1, 4, 5, 6 and 9 (though it is also possible for a number to end with one of these digits without being a square).

Iwaniec and Friedlander used a sieve for primes based on the form $a^2 + b^4$ and showed that among numbers of that form there are infinitely many primes. For example, $1^2 + 2^4$ is 17, which is a prime, though $2^2 + 3^4$ is 85, which is not. This was a significant advance because in numbers up to, say, a trillion (1,000,000,000,000), there are fewer than a billion (1,000,000,000) numbers that can be written in the form $a^2 + b^4$, which means that you can 'sieve out' a bunch of numbers with a higher proportion of primes in it than had been possible previously.

The story of Iwaniec's relationship with the Riemann Hypothesis began very early in his life. 'I was in Poland, and I was exposed to mathematics early because of the Mathematical Olympiads [an international mathematics competition for schoolchildren]. I don't say that we had a fantastic programme in school, but I was competing in mathematics very early and at the same time learning mathematics much on my own rather than from school. And I had a friend who loved to write letters. I couldn't keep up with him because he was writing ten letters a week, and I responded with one. And at one point he asked me, "Henryk, do you know what is the Riemann zeta function?" I answered, saying, "I'm sorry, I don't know what it is at all," and he said, 'But I heard this is so good, it is something to do with prime numbers, and there's a great conjecture. Can you please find something, where it is, what it is?" So I went to my teacher and asked him what is the zeta function. The next day the teacher called me and explained me everything, saying, "This is the zeta function." Now that I understand what he said, he actually gave me the *Weierstrass* zeta function, not the Riemann zeta function, but in the meantime my friend wrote and sent me copy of Riemann's memoir, so I was really exposed to it early. I was in my third year in high school, aged seventeen.'

Iwaniec's favourite topic at that age was what is called complex analysis, the mathematics of the number system that incorporates numbers of the form $a + ib$, where i is the square root of -1. His early passion for mathematics led him to believe, when he was an undergraduate, that he had a proof of another important theorem.

'There was a famous professor in the Academy of Sciences who had a seminar there, and he worked in number theory. One day, because of my interest in prime numbers, I believed that I had a proof of the Dirichlet theorem for primes in arithmetic progressions, and so I said to one of my professors in the department that I think I have a proof.' The professor spoke to another teacher, Professor Schnitzel, and arranged for Iwaniec to see him. 'Next day, I was embarrassed because I found a mistake in my calculations. I said, "No, I'm not coming, no way, no way", and then he said, "No, please come nevertheless," so I went. Professor Schnitzel was extremely tactful and never asked me a question about what I had done before, he simply gave me a paper to present in the seminar. I was thrilled – this was the Institute of Mathematics at the Polish Academy of Sciences, and I was just a first-year undergraduate student. The famous professor was very tactful not to comment on what I had claimed days before. The paper he gave me was [Enrico] Bombieri's paper on an elementary proof of the Prime Number Theorem, a very, very advanced paper. I suffered. I read this paper day and night for two weeks before I presented it. And I loved the subject so much that even today it's inevitably the Riemann Hypothesis that I'm working on because of prime numbers.'

Enrico Bombieri is now at the Institute for Advanced Study at Princeton. People warned me before I went to see him that I'd have to sit quite close to him and listen carefully, as he talks softly and slowly. Like Iwaniec and many of the top mathematicians in this field, Bombieri was gripped at an early age by prime numbers and the Riemann Hypothesis.

'At the age of eight or nine,' he told me quietly, 'I came across some books on geometry and algebra, and a book on mathematical problems. Number theory, because it required less foundational materials, attracted my attention. My father was interested also in mathematics. He was a banker, but maybe he wanted to be a mathematician and from time to time he bought books on mathematics and tried to read them. So by the age of twelve or so I was helping him, and bit by bit, as books became part of my personal library, I would ask my father, "Can you find this book for me?" And so he would find it.

'Then I started thinking about problems and even doing the beginnings of research, and my father thought it was a good idea to get some professional opinion on what I was doing. Eventually I had a little manuscript which was examined by Giovanni Ricci, a professor at University of Milan and at the time the leading Italian specialist in number theory, and he read the manuscript – he didn't know it was written by a teenager, I was fifteen – and when he read it he said, "Who is this guy? I want to get him as my assistant." And then it was explained he could not do that because I was still in high school, but I went to see him and so he started coaching me.

'I knew quite early about the Riemann Hypothesis, that it was really the key to understanding the finer distribution of prime numbers, and I was interested in prime numbers and analytic number theory and so it was natural to come across the Riemann Hypothesis. By the time I was eighteen I was reading everything, and bit by bit I got interested in these things.'

Nowadays, Bombieri is one of the acknowledged leaders in the Riemann Hypothesis field.

Unless you've had close contact with an eight-year-old who has asked you for an algebra book for Christmas, it's probably difficult to look into a child's mind see just what it is that leads to a preference for mathematics over any other activity. Yet, as I spoke to mathematician after mathematician about their childhoods, the stories poured out. Granted, it's not too difficult for an inexperienced mind to grasp the concepts of number theory and some of the basic questions, but it's still extraordinary how for some this childhood interest has become a lifelong preoccupation.

Matti Jutila, a Finnish mathematician, is another. When I was talking to him in his office at the University of Turku, he suddenly leaned back and took from his bookshelf a slim and slightly battered volume, and began to read:

'"*Niiden alkulukujen lukumäärää, jotka ovat ≤ n...*" This is where my interest in number theory started. This is a Finnish textbook on number theory and algebra. It was a university textbook. I bought this when I was sixteen years old. I bought it in the spring; I read it through during the summer without any guidance from anybody – just by myself. I had only an elementary knowledge of number theory at the time but I managed when I entered the university to pass an exam based on this without participating in the course.'

He pointed to a two-line footnote, where the only recognizable thing to a non-Finnish speaking mathematician was the formula giving Gauss's estimate for the number of primes less than any number n.

'These two lines determined my choice of topic. It is here at the very moment that, with these two lines, I decided that this was to be my field. Behind the decision was a fascination with the prime numbers...The Riemann Hypothesis is a fundamental problem. This is a first-class problem. Compared for instance with Fermat's Last Theorem, I have always felt that Riemann's Hypothesis is genuinely important. An interesting problem which leads to something if it is solved or proved. Who knows? Theoretically, it is of course possible that the hypothesis is false. Most people believe the contrary.'

Most *people*, in fact, have no view on the matter either way. Even as a non-mathematical reader having reached this far, you still know very little about the Riemann Hypothesis. You know that there is a mathematical expression that predicts roughly how many prime numbers there are smaller than any number you care to name, if you put that number into the expression. You know also that this prediction, by Gauss, is not entirely accurate, and that the amount by which it is wrong is the subject of another mathematical expression, devised by the German mathematician Bernhard Riemann. With Gauss's estimate, proved by two other mathematicians in 1896, and Riemann's correction, conjectured but not yet proved by anyone, we know much more about how the prime numbers are distributed.

But there's still an ingredient that I have glossed over so far. At the heart of Riemann's correction factor, and essential to understanding how it is related to prime numbers, is Riemann's zeta function and, in particular, a series of numbers which are known as the Riemann zeros – the 'zeros of zeta of s', in Tom Apostol's lines of verse quoted at the start of this chapter.

The California Institute of Technology – Caltech – sprawls over several acres of Pasadena, a satellite city of Los Angeles. When Caltech sought to expand its elegant neoclassical campus by buying adjacent land, the city fathers allowed them to do so provided they didn't knock down the many large houses that lined several of the nearby suburban streets. As a result, finding Tom Apostol in his office meant meandering down a wide tree-lined avenue, past spurting lawn sprinklers, and peering across well-clipped grass to read the discreet signs placed next to solid domestic front doors.

Apostol was in a house bearing the name 'Project Mathematics!'. I found him upstairs in what was once a bedroom, staring intently a screen full of Greek. From pi onwards, maths has made extensive use of Greek letters, but what Apostol was looking at wasn't maths. It was the text of a speech he had to make in Greek in a few weeks' time. Apostol's parents came from Greece to America, and as a child he spoke the language but had never learnt to read or write it properly. Now, as he did from time to time, he was returning to the land of his ancestors, and on this occasion he'd decided to risk giving a lecture in Greek.

Apostol – originally 'Apostolopoulos' – is a tall, grizzled man in his seventies, with short white hair and a craggy face. Although he may never prove the Riemann Hypothesis, his teaching has spawned three generations of mathematicians, any one of whom might. Mathematicians can be divided into two types, both necessary for a successful proof: *explorers*, who map the likely working area, and *diggers*, who do the hard graft necessary to find the final proof. There's also a third type, essential to success – the *guide*, who inspires the explorers and diggers by first helping them to understand the problem and why it is important and interesting. Of the mathematicians I met who were working on the Riemann Hypothesis or the Riemann zeta function, many were related to one another on a pedagogical 'family tree'. More than one had been taught by Brian Conrey, Conrey had been taught by Charles Ryavec, and Ryavec had been taught by Tom Apostol.

Apostol is a guide who seems to have the knack of producing the diggers and the explorers. He told me of a remark which shows the kind of impression he can make.

'One of my students once came up to me after one of my calculus lectures and says "Dr Apostol," he says, "these things seem to *mean* something to you – you seem so at home with all this stuff." '

Although he does not feel he has the specialized knowledge or the stamina to prove the Riemann Hypothesis, he nevertheless keeps returning to it, like the tip of a tongue returning to a cavity in a tooth. I first came across him when I discovered a poem that began, 'Where are the zeros of zeta of *s*? / G.F.B. Riemann has made a good guess…', and I tracked him down. 'Poem' is perhaps too elevated a word for a *jeux d'esprit* written in 1955 that occasionally violates scansion by cramming several extra syllables into a line in order to get in names such as Landau, Bohr, Cramér, Littlewood and Titchmarsh. But there are also felicitous couplets such as 'Related to this is another enigma / Concerning the Lindelöf function μ(σ)' (read this last expression 'mew of sigma'). But the verses, originally composed as an after-dinner turn at a conference of number theorists, have somehow survived in the informal annals of mathematics, and people occasionally add new verses to bring the story up to date.

Apostol has been in the world of academic mathematics for fifty years. He originally intended to be a chemical engineer but, like so many others, got the bug for pure maths when he was given an unsolved problem by his professor which he managed to solve. The knowledge that he had a made a discovery – albeit a minor one – that no one else had ever made was enough to change his career path from chemical engineering to maths. The problem he solved was in the area of magic squares, usually seen merely as an intriguing puzzle for those who like mathematical recreations. But as often happens in number theory, even the simplest questions can lead to complicated mathematics. Writing the numbers from 1 to 25 in a 5×5 grid so that the rows, columns and diagonals all add up to the same number is something that anyone can do by trial and error. But if you make the grid much bigger, trial and error becomes time-consuming. In any case, a mathematician would be more interested in trying to find a rule that will tell you how to fill in the squares for any grid, however big. It was working on a version of that rule that made Apostol feel that he wanted to be a mathematician.

Like all pure mathematicians, he was familiar with the Riemann Hypothesis but he never saw himself as a serious contender.

'I had no idea how to approach it,' he said. 'I had read all the stuff and realized it was a very difficult problem. Some of us used to talk about it, and I remember once somebody said, "I had a dream one night – take zeta of zeta of s." Well, I never tried to push that.'

When mathematicians start dreaming about a particular mathematical function, it's clear it has it has begun to take over their lives. With the Riemann zeta function, part of its fascination lies in its complexity, in both senses of the word. It is very complex to understand because, as we've seen, it uses complex numbers where the simpler Euler zeta function uses real ones.

Both zeta functions are described by the expression $\zeta(s)$, but in the Riemann zeta function s is a complex number, a number of the form $a+ib$, where i stands for the square root of minus 1. The letter z in mathematical expressions is usually taken to mean a *complex* number, whereas, on the whole, x and y are used for *real* numbers. So the Riemann zeta function is the sum of the series

$$1 + \frac{1}{2^z} + \frac{1}{3^z} + \frac{1}{4^z} + \frac{1}{5^z} + \cdots$$

which means

$$1 + \frac{1}{2^{a+ib}} + \frac{1}{3^{a+ib}} + \frac{1}{4^{a+ib}} + \frac{1}{5^{a+ib}} + \cdots$$

The sum of this series of fractions can take on all sorts of values, depending on what number is substituted for z (or, in other words, a and b). But the entire army of explorers and diggers working on the Riemann Hypothesis are only interested in one value for the sum of the Riemann zeta function. They are only interested in the function when all the terms of the series add up to zero. (Bear in mind here that just because the series of terms has plus signs, it doesn't mean that it grows and grows, never equalling zero. If some of the terms are negative, if $1/n^{a+ib}$ is sometimes negative, for example, then they could cancel out and make the sum of the series zero.)

There is one huge assumption here which has to be swallowed by anyone who is not entirely familiar with 'functions of a complex variable' (which includes me). It is not revealed to the non-mathematical human mind how on earth the expression $1/n^{a+ib}$ can have any numerical significance. To put it bluntly, with an actual example, say, $1/7^{3+12i}$, how do you raise 7 to the power '3 + 12i'? We might be able to move it one stage further, if we remember that 7^{3+12i} is the same as $7^3 \times 7^{12i}$, but at this point if we really want to understand how mathematicians raise a number to an imaginary power and then divide it into 1, we need to put this book down, leave the house, sign up for a few months of complex analysis and number theory, and then pick up the book again in a year or two.

But if you'd rather read on – as I hope you will – you must take it on trust that some (complex) numbers of the form $a+ib$, when they are substituted for s in the Riemann zeta function, make the value of the function zero. The first part of the expression, the a, is called the real part and the second, the b, is called the imaginary part, and here are the first few of those numbers:

$$\frac{1}{2} + 14.135i, \quad \frac{1}{2} + 21.022i, \quad \frac{1}{2} + 25.011i, \quad \frac{1}{2} + 30.425i,$$

$$\frac{1}{2} + 32.935i, \quad \frac{1}{2} + 37.586i$$

These are the points where the simplified graph of the Riemann zeta function shown in Figure 5 meets the zero axis. The rule that determines

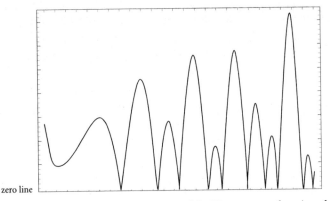

FIGURE 5 In this simplified representation of the Riemann zeta function, the points where the curve of the function dips to zero represent Dr Riemann's zeros.

where those zeros are is as mysterious as the rule that governs the distribution of the primes. The two sets of numbers – the zeros and the primes – are inextricably linked in a way which we will only understand when someone proves the Riemann Hypothesis.

There are certain facts about the Riemann zeros that we suspect but can't prove, and others that we know because they have been proved. The first few zeros all definitely have the real part equal to one-half, and we suspect that *all* Riemann zeros have the same form, $\frac{1}{2} + it$, where t is some real number – but we can't prove it. What we *can* prove – indeed Riemann himself proved – is that all the zeros have their real part somewhere between 0 and 1.

For Charles Ryavec, the Riemann zeta function has all the untranslatability of a piece of subtle English verse into a foreign language. He writes:

Hilaire Belloc's pun

> When I am dead, I hope it may be said:
> His sins were scarlet, but his books were read.

comes to mind, and I wonder how much the French reader (say) ignorant of English, can appreciate it...The function $\zeta(s)$ embodies a considerable condensation of meaning, with just about everything lost in the literal translation: 'the sum of the reciprocals of the natural numbers raised to the complex power, s.'[57]

Some mathematicians imagine the Riemann zeta function, as they do other functions in mathematics, as a mathematical 'object'. They understand it so well that in their minds it assumes a shape or a structure, but this structure is not really morphological – it doesn't have a top or a bottom, it's not coloured or textured, nor does it have a weight. Nevertheless, it has parts which relate to one another, it can be transformed from one 'shape' into another, and it can be probed by mathematical tools which reveal hitherto unknown aspects of its 'internal' arrangements. All these mental images combine to give a mathematical function the key characteristic of a structure – its stability and longevity.

As non-mathematicians trying to understand the Riemann zeta function better, we have to externalize it, to winkle it out of the brains of the mathematicians. We've already done that in a very simple way, trying to represent the fluctuations of the function as a curve that repeatedly shoots up from the zero line and back again. But this is such a long way from how the function really behaves that it's worth trying to get closer to it with a more complicated description, based on the representation of mathematical functions by three-dimensional graphs. (For more on graphs in maths, see *Toolkit 5*.)

To recap: with the function (or 'black box') $\zeta(s)$, if you put a value of s in at one end, you get a value $\zeta(s)$ out at the other end. With graphs of conventional functions using real numbers, mathematicians plot x against y, for example, and see how the function turns a series of increasing values of x into some shape – a slanting line, an ellipse, a parabola. Because s is a complex number, we can't use exactly the same method but have to find an analogous way of providing a useful visual representation of the zeta function.

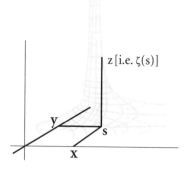

FIGURE 6 We can visualize the Riemann zeta function as a three-dimensional curve if, for every point s with coordinates x and y on the horizontal plane, we calculate a point z at some height above the plane to represent $\zeta(s)$. If we do this for every point in the x,y plane we will map out a three-dimensional surface.

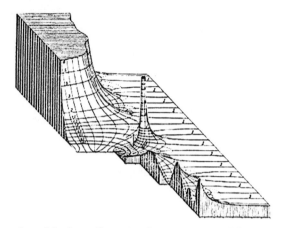

FIGURE 7 A section of the three-dimensional representation of the Riemann zeta function. The critical line is shown slicing through the three nearest 'mountain peaks'. The tall peak in the middle actually rises to infinity because when s = 1, $\zeta(s)$ is infinitely large.

Since s is complex, we can represent it as $x + iy$, and so in Figure 6 s is shown as a point with two coordinates on the horizontal plane. If we use our streets analogy, s is a point, at address y, along the street that runs off Camino Real, leaving it x units from its origin. Now, we need a way to show graphically what happens to each value of s if we 'process' it using the zeta function. Although the answer – call it z – is also a complex number, there is a way to represent it as a simple number, a 'height' above the horizontal surface. So for each point s there is a corresponding point z – which equals $\zeta(s)$ – located somewhere in three-dimensional space. If we take a series of points, $s_1, s_2, s_3, s_4, \ldots$, and work out $z_1, z_2, z_3, z_4, \ldots$, we will see the outlines of a three-dimensional shape begin to emerge, as shown in Figure 6.

So now we have a way of representing a relationship between complex numbers as a curved surface in three-dimensional space. If we do this for just a small portion of the Riemann zeta function, we get a surface whose topography has much to tell us. Figure 7 shows a slab of space around the most interesting area of the Riemann zeta function, where x is between –6 and +8 and y goes from +26 to –26. In this type of graph, the values of the zeta function for $s = x + iy$ are the heights of the surface above zero (which we could call sea level).

The peak in the middle of Figure 7 is a 'mountain' which rises all the way up to infinity. It's the value of $\zeta(s)$ where $s = 1$. We can see that by just looking at the series for $\zeta(s)$:

$$\sum \frac{1}{n^1} \quad 1 + \frac{1}{2^1} + \frac{1}{3^1} + \frac{1}{4^1} + \frac{1}{5^1} + \cdots$$

Or in other words, $1 + \frac{1}{2} + \frac{1}{3} + \frac{1}{4} + \frac{1}{5} + \dots$. This is a famous series called the harmonic series, and you can make it grow as large as you like by adding more terms, even though the terms themselves get smaller. So an infinite number of terms gives us an infinite value for $\zeta(1)$. Such a point where a surface shoots off to infinity for some value of a function is called a pole.

But where do the zeros fit into all of this? As those values of s that make $\zeta(s)$ equal zero, they are the places on the 'landscape' where the land reaches sea-level. In Figure 7 the slab of land with the nearest 'mountains' is sliced through along a straight line and we can see there are three points at which the valleys between the mountains dip down to sea level. These three points, the first three zeros, are in a straight line. This line is half a unit away from the zero axis on the horizontal plane. And so every point on the line has the form $\frac{1}{2} + iy$. There are more zeros in this landscape, stretching away into the distance, and all the zeros that have ever been calculated have been found to lie on this line. For this reason it's called the critical line. But no one has yet proved this. The nearest mathematicians have got is to prove that Riemann zeros *must* lie somewhere in a strip that is half a unit either side of the critical line, known as the critical strip.

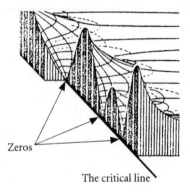

FIGURE 8 A closer view of the three nearest 'mountain peaks' of the Riemann zeta function surface shown in Figure 7. The first three Riemann zeros are shown.

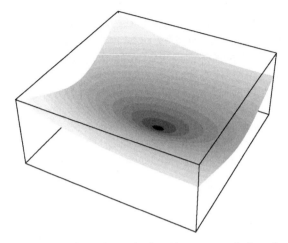

FIGURE 9 Seen obliquely from above, the first Riemann zero is the point where the three-dimensional surface of the Riemann zeta function touches the horizontal x–y plane, at the bottom of the 'box'. This point is at the centre of the darkest part of the surface.

Figure 9 shows an intimate portrait of just one of the Riemann zeta function zeros. Seen from above, the sloping sides get darker as the 'land' gets nearer to 'sea level'. At just one point, in the middle of the darkest part of the dip, is where a Riemann zero will be found, at the value of s for which the Riemann zeta function equals zero. This particular section of the Riemann zeta curve shows the area around the smallest Riemann zero.

Now that powerful computers are widely available, you can calculate as many Riemann zeros as you wish, but from Riemann's time until the 1960s, such calculations had to be done by hand or with mechanical calculators, and it was an arduous process. For sixty years after Riemann's death it was believed that he hadn't actually calculated any zeros at all, but based his hypothesis on a mixture of reasoning and hunch. Sir Michael Berry is closely connected with one of the most promising approaches to a proof. He is a genial mathematical physicist at the University of Bristol who wears his knighthood lightly, and he told me how it was discovered that Riemann actually had calculated some of the zeros:

It's a very romantic story. Riemann was a very disorderly worker, and when he died he left masses of scribbles. It was sixty years before someone, the very great mathematical dynamicist Carl Ludwig Siegel, looked at Riemann's papers more carefully. He found scribbled there some formulas he couldn't understand, overlaid with weird things like calculating the square root of 2 to 38 decimal places, and he worked at them and realized that Riemann had in his hands a way of calculating the Riemann zeros, a beautiful piece of mathematics, which if Siegel hadn't rediscovered it we probably wouldn't know about even now. So Siegel dug all this out and put it into a decent form, and it's called the Riemann–Siegel formula, and almost everyone uses this to compute the high zeros. It's fabulous.

Calculations in analytic number theory are rarely straightforward. They often involve representing a brief expression containing complex numbers as a combination of sums of infinite series in which the numbers are real and therefore can be manipulated. But once a manageable method, the Riemann–Siegel formula, became available for calculating the zeros of the zeta function, the numbers of known zeros of the Riemann zeta function increased from half a dozen or so to several dozen in the next thirty years. In 1935, an Oxford mathematician, E.C. Titchmarsh, used a punched-card calculating device to find the first 104 zeros. Then, with more sophisticated equipment, many hundreds and soon thousands could be calculated. And every single zero was of the form $\frac{1}{2} + ib$.

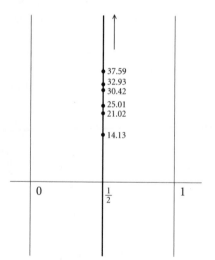

FIGURE 10 The first few Riemann zeros to be calculated were all of the form $\frac{1}{2} + ib$, and the value of b for the first zero is 14.13. It can be proved that all zeros lie somewhere in the critical strip, but no zero has yet been discovered that is not on the critical line running down the middle of the critical strip.

What this means, if we go back to the simpler representation of complex numbers as addresses on a grid of streets, is that all the values of the zeros of the Riemann zeta function found so far lie on a single street, at right angles to Camino Real, beginning at a point very near the beginning of the real numbers, the point $\frac{1}{2}$, as shown in Figure 10. Now, what Riemann said – without any proof – is that *all* the zeros* that can ever be calculated will lie on the same 'street', with street 'name' $\frac{1}{2}$, stretching away from the real number axis from the point $a = \frac{1}{2}$. And this line, which will assume greater importance as we get even deeper into the topic, is the critical line.

In fact, he didn't quite say that, as Alexander Ivic pointed out to me. 'Riemann in 1859 actually said, "*sehr warscheinlich*" – meaning "very probably" – "all the zeros are on the critical line". Now, he was a pioneer, a visionary. His visions and results were several decades ahead of his contemporaries, but note that he was cautious – he said, "very probably" – he did not say, "I would bet my life that this is true".'

It's now possible to understand a little of Tom Apostol's verses, armed with a few of the basic elements. Apostol told me how he came to write them, prompted by a first line supplied by someone else.

'Somebody said to me once, "Hey, have you heard that poem about the zeta function?" I said, "No. How does it go?" He says, "The first line is, 'Where are the zeros of zeta of s?'" "Yes?" "That's all. That's all I've got." That triggered my interest and I said, "How could you make a poem out of that? – 'Where are the zeros of zeta of s? G.F.B. Riemann has made a good guess.'" And we were going to have this number theory conference at Caltech, and I decided to write something to be performed at a party we were going to have. A few versions were written, using the music from "Sweet Betsy from Pike". It was a big hit.'

Here's a selection of the more accessible (or less inaccessible) verses:

Where are the zeros of zeta of s?
G.F.B. Riemann has made a good guess,
They're all on the critical line, said he,
And their density's one over $2\pi \log t$.

---

* Apart from what are called 'trivial zeros', which have no real relationship to the prime numbers.

This statement of Riemann has been like a trigger,
And many good men, with vim and with vigor
Have attempted to find, with mathematical rigor
What happens to zeta as mod *t* gets bigger.

The names of Landau and Bohr and Cramér,
And Hardy and Littlewood and Titchmarsh are there,
In spite of their efforts and skill and finesse,
In locating the zeros no one's had success.

In 1914 G.H. Hardy did find
An infinite number that lay on the line,
His theorem, however, won't rule out the case
That there might be a zero at some other place.

Oh, where are the zeros of zeta of *s*?
We must know exactly. It won't do to guess.
In order to strengthen the prime-number theorem,
The path of integration must not get too near 'em.

Over the years since Riemann first published his famous paper, 'many good men, with vim and with vigor' have tried to prove the Riemann Hypothesis. Quite a lot of them have even announced that they had a proof. But almost inevitably, as the history of the topic shows, one by one their proofs have been demolished as one or more of their colleagues took on the painstaking task of finding the errors they knew were almost certainly there.

6　Proofs and refutations

Few understand it, none has proved it.

Time magazine, 30 April 1943

There have been several occasions in the past hundred years when mathematicians have announced that they have proved the Riemann Hypothesis. But none can have been stranger than the events that took place in New York early in 1959,[58] at a lecture at Columbia University sponsored by the American Mathematical Society. A brilliant thirty-year-old mathematician, John Nash, proposed to announce a proof of the Riemann Hypothesis. The story of Nash's life – his early brilliance, his descent into madness and his remarkable, though fragile recovery – has been told in Sylvia Nasar's book *A Beautiful Mind*, and, in a glamorized form, in the feature film of the same name.

Since the age of fourteen, Nash had been captivated by the Hypothesis after reading about it in E.T. Bell's classic work, *Men of Mathematics*. Nash had been hailed by *Fortune* magazine as 'the most promising mathematician in the world', so he was certainly considered by his colleagues as capable of producing a proof. He had also confided to some friends and colleagues that he believed he had an idea that might work, using what are called pseudoprimes, whole numbers that are not primes but behave like them in certain ways.

So when two hundred and fifty or so mathematicians gathered to listen to Nash, there was an air of expectation in the lecture hall. Perhaps he really had done it? But as Nash got deeper and deeper into his 'proof', the audience of mathematicians became more and more aghast. It wasn't just wrong – it was nonsense.

According to Donald Newman, 'One word didn't fit in with the other. Everybody knew something was wrong. It was his chatter. The maths was just lunatic. What does this have to do with the Riemann Hypothesis? Some people didn't catch it. People go to these meetings and sit through lectures. Then they go out in the hall, buttonhole other people, and try to figure out what they just heard. Nash's talk wasn't good or bad. It was horrible.'[59]

Cathleen Morawetz, a friend of Nash's, ran into him in the hall afterwards. 'He was laughed out of the auditorium', she said. 'I felt terrible. I said something nice to him but I was disturbed. He seemed very depressed.'

But some were less surprised than others. A month previously, Nash had been talking to a friend and told him that he, Nash, was on the cover of *Life* magazine, disguised as Pope John XXIII. 'How do you know that it is you if it looks like the pope?' asked the friend. Nash replied that, first, it was because John wasn't the pope's given name – he had chosen it. And second, 23 was Nash's favourite prime number.

In fact, the lecture was a florid manifestation of Nash's schizophrenia, which had been developing for some time and was eventually to incapacitate him for many years. By the 1990s, though, he was on the mend, and in 1994 he was able to go to Stockholm to receive a Nobel Prize for Economics, for work he had done many years before.

No one can say whether it is significant that Nash had been focusing on the most difficult problem in mathematics in the time leading up to his madness. But Nash himself certainly believed that mathematicians showed a propensity to go mad. 'I would not dare to say that there is a direct relation between mathematics and madness,' Nash said in 1996, 'but there is no doubt that great mathematicians suffer from maniacal characteristics, delirium and symptoms of schizophrenia.'

There is no greater indication of the difficulty and importance of Riemann's Hypothesis and his zeta function than the roll call of distinguished mathematicians who have tried and failed to prove it. H.M. Edwards lists several in his book about the Hypothesis:

> No mathematical work is more clearly a classic than Riemann's memoir *Über die Anzahl der Primzahlen unter einer gegebenen Grösse*, published in 1859. Much of the work of many of the great mathematicians since Riemann – men like Hadamard, von Mangold, de la Vallée-Poussin, Landau, Hardy, Littlewood, Siegel, Pólya, Jensen, Lindelöf, Bohr, Selberg, Artin, Hecke, to name just a few of the most important – has stemmed directly from the ideas contained in this eight-page paper. According to legend, the person who acquired the copy of Riemann's collected works from the library of Adolf Hurwitz after Hurwitz's death found that

the book would automatically fall open to the page on which the Riemann Hypothesis was stated.[60]

Even Alan Turing, the British mathematician who played such an important part in the British deciphering operation at Bletchley Park during the Second World War, was seduced by the fascination of the Riemann Hypothesis. In the midst of laying the theoretical foundations of what were to become digital computers, Turing designed a machine to calculate zeros of the Riemann zeta function. Turing's biographer, Andrew Hodge, writes:

> Apparently [Turing] had decided that the Riemann Hypothesis was probably false, if only because such great efforts have failed to prove it. Its falsity would mean that the zeta function did take the value zero at some point which was off the special line, in which case this point could be located by brute force, just by calculating enough values of the zeta function.[61]

Turing did much of his own engineering work, planning to construct a system of eighty meshing gearwheels with weights attached at specific distances from their centres. As the wheels rotated, the combined effect of different weights rotating at different distances from their axes would produce a varying turning effect – a moment, as it is called – and that moment, Turing hoped, could be estimated by balancing the whole system with a counterweight. From the magnitude of this moment, after each set of rotations of the wheels, a new Riemann zero could be calculated. At least that was the theory.

Turing's friends got used to the resulting chaos of his room in King's College, Cambridge, where he had been made a Fellow in 1935:

> It was liable to be found with the sort of jigsaw puzzle of gear wheels across the floor. Kenneth Harrison, now a Fellow, was invited in for a drink and found it in this state. Alan tried and lamentably failed to explain what it was all for. It was certainly far from obvious that the motion of these wheels would say anything about the regularity with which the prime numbers thinned out, in their billions of billions up to infinity. Alan made a start on doing the actual gear cutting, humping the blanks along to the engineering department in a rucksack, and spurning an offer of help from a research student. Champ [a friend, David Champernowne] lent a hand in grinding some of the wheels, which were kept in a suitcase in Alan's room.[62]

But the machine was never finished, as more pressing matters inter-vened – the Second World War and the need to crack the enemy's enciphered messages. There's a small irony in the fact that, today, number theorists working with prime numbers are in the frontline of code-making and code-breaking activities around the world, and knowledge gained from attempts to prove the Riemann Hypothesis could lead to virtually unbreakable codes.

What Turing was trying to do was to use his machine to find at least one zero that didn't fit the Riemann Hypothesis, a zero that wasn't of the form $\frac{1}{2} + ib$, but started with some other real number. We now know that even if he had successfully assembled his contraption of strings and brass and started cranking away in his rooms in Cambridge he would never have been able to go through enough zeros to find one that disproved the Riemann Hypothesis. Modern computers – whose intel-lectual origins can be traced back to Turing – have shown that many billions of zeros all lie firmly on the line with real part equal to a half, and none has been found off the line.

But counting is not proving. What the world needs is proofs. And it's not as if there haven't been proofs. In 1896, when Hadamard pub-lished his proof of the Prime Number Theorem, he apologized for the meagreness of his contribution to the subject, in the light of 'the recent proof of the Riemann Hypothesis claimed by Thomas Stieltjes'.[63] Needless to say, Stieltjes found a gap in his own proof after he made the claim and he had to retract. It was not the first occasion, and would not be the last, on which a distinguished mathematician believed he had within his grasp the Holy Grail of number theory. Such an occasion even reached the pages of *Time* magazine, in its issue of 30 April 1943.

Under a photo of Riemann with the caption 'Few understand it, none has proved it' the *Time* reporter wrote:

A sure way for any mathematician to achieve immortal fame would be to prove or disprove the Riemann hypothesis... No layman has ever been able to understand it and no mathematician has ever proved it.

One day last month electrifying news arrived at the University of Chicago office of Dr. Adrian A. Albert, editor of the *Transactions* of the American Mathematical Society. A wire from the society's secretary, University of Pennsylvania professor John R. Kline, asked editor Albert to stop the presses; a paper disproving the Riemann Hypothesis was on the way. Its author: Professor Hans Adolf Rademacher, a refugee German mathematician now at Penn.

On the heels of the telegram came a letter from Professor Radamacher himself reporting that his calculations had been checked and confirmed by famed mathematician Carl Siegel of Princeton's Institute for Advanced Study. Editor Albert got ready to publish the historic paper in the May issue. US mathematicians, hearing the wildfire rumor, held their breath. Alas for drama, last week the issue went to press without the Rademacher article. At the last moment the professor wired meekly that it was all a mistake; on rechecking, mathematician Siegel had discovered a flaw (undisclosed) in the Rademacher reasoning. US mathematicians felt much like the morning after a phoney armistice celebration. Said editor Albert, 'The whole thing certainly raised a lot of false hopes.'

Functions with complex numbers require careful handling since you cannot assume that they behave in the same way as functions with real numbers. For example, there are only two numbers that can be roots of 1 in the real number system: +1 and −1. The square roots of 1 are −1 and +1; the cube root is −1; the fourth roots are −1 and +1, and so on. But in the complex field, 1 can have any number of different roots: twenty-nine 29th roots, for example, all different.

The mistake in Rademacher's proof centred on the fact that logarithms of complex numbers behave very differently from logarithms of real numbers.

'The complex logarithm does not have a uniquely defined value,' said Samuel Patterson, 'and Rademacher chose two different values at two different points in the argument. It's very easy to get a contradiction if you do that – if you assume the Riemann Hypothesis is false and then do this, this, and this, and you get a contradiction, then it must be true. Professor Cassels'* comment in his lectures was – he would say, "So we will prove this by contradiction – we do this, this, and this and then we've got a contradiction. This shows that either our original assumption was wrong or we have made a mistake."'

Even the best mathematicians can make mistakes, as Brian Conrey explained: 'Just like anybody, you can be blindsided. When you're working on something, if you think you've got a proof, especially if it's really good, you immediately become blind to the flaws and it gets very hard

---

* John Cassels, a famous English number theorist, at Trinity College, Cambridge.

to see the mistake. You go back over it and you go back over it, but you just keep skipping over that place. You often have to show it to somebody else to try and get them to check through it, and it's easy to bury a mistake somewhere in the middle of some long paper. I don't think people really do it intentionally, but you want it so much it's just easy to overlook and be blinded.'

The mathematical community gets very jittery when it comes to proofs of the Riemann Hypothesis. Conrey gave a couple of examples of how people are only too ready to believe rumours. One rumour concerned a famous mathematician called Norm Levinson.

'In 1974 Levinson discovered a new method by which to prove that a positive proportion of the zeros were on the critical line', said Conrey. 'The calculation was brutally technical, but Levinson thought he could prove that 98.6 per cent of the zeros were on the critical line. He showed Gian-Carlo Rota a manuscript one morning and said that he could prove that 100 per cent of the zeros of the zeta function were on the critical line. Rota said, "Really?" Levinson held up the manuscript and jokingly replied, "Yes, in here I prove that 98.6 per cent of the zeros are on the line. I leave the other 1.4 per cent to the reader." Rota, not understanding the joke, and not realizing that 100 per cent of the zeros on the line is not the same as the Riemann Hypothesis, started telling people that Levinson had proved the Riemann Hypothesis. By the afternoon the news had reached the West Coast. It was before e-mail, but news travels fast. There was a mistake discovered in Levinson's result, but he really had proven that at least one-third of the zeros are on the critical line. Modifications of his method led to the current record of 40 per cent of the zeros on the line.'

It's another of those surprises that maths springs on the unwary to discover from Conrey's statement that '100 per cent of the zeros on the line is not the same as the Riemann Hypothesis'. In other words, 100 per cent is not the same as 'all'.

Andrew Granville tried to explain why. 'What we mean in number theory,' he began, 'is if you take the numbers up to 100 and say that 90 have a certain property, that's 90 per cent, and then you look up to 1,000 and 990 have the property, so that's 99 per cent, and you look up to 10,000 and you get 99.9 per cent, and you keep on going. Now at no time was it that *all* the integers have this property, but the percentage is gradually increasing, and as you go up to infinity it's going to go up to 100

per cent. And that's what we mean by 100 per cent, and why it's not "all".
And there is a big distinction. In fact I proved in 1985 that Fermat's Last
Theorem is true for 100 per cent of exponents – so with $x^n + y^n = z^n$, for
100 per cent of n there are no solutions. But it was an absolutely meaningless result. There's no way you could develop this and say "for all
exponents", and it said nothing profound.'

Granville described how Atle Selberg had written a paper saying that
the probability was 100 per cent that in short intervals the number of
primes is exactly as you would predict. Then a mathematician called
Helmut Meier wrote a paper which destroyed the argument.

'Reading that paper is just a delight,' said Granville. 'There are hundreds of thousands of papers written of no conceivable interest, but
Helmut in four pages strung together some ideas in an unbelievably
original way and showed that everybody was wrong, even Selberg.
Everybody was wrong.' What Meier showed was that, with a proposition
where the probability is 100 per cent, there can still be an infinite
number of counter-examples. 'That's what was so shocking, that there's
loads of counter-examples, quite a big infinite number in some technical
sense.'

So great is the desire of mathematicians to see a proof of the Riemann
Hypothesis that they can easily be fooled in their eagerness.

'There was an April 1st joke that went around,' Brian Conrey told me.
'Alain Connes has outlined a program that could conceivably prove the
Riemann Hypothesis, and he was giving some lectures at the Institute
[for Advanced Study]. An e-mail went around on April 1st – it was very
funny – saying that a bright physics student in the audience listening to
Connes recognized something about bosons and fermions,* and saw
the right way to do this and so completed the proof that evening. That
circulated, and lot of people thought it was true. But it was an April
Fool's joke.'

In addition to well-known mathematicians' attempts to prove the
Riemann Hypothesis, there are plenty of cranks who believe they have
proved it – people with a half-baked understanding of the subject who
don't appreciate the subtleties of functions of a complex variable.

Roger Heath-Brown has a lot of experience with these people. 'I
receive unsolicited manuscripts quite frequently – one finds a particular

* Subatomic particles.

person who has an idea, and no matter how many times you point out a mistake they correct that and produce something else that is also a mistake. One category would be people, often in third-world countries, for example, who have a university position and have a lifetime to devote to mathematical research but perhaps don't have the same formal training that an academic in Britain does. But equally there are some top-rate mathematicians of the more erratic type who keep plugging away at this, and particularly with them it can be very difficult to locate an error. One of the things that concerns me is why I should have to devote a week of my time searching through a manuscript which I'm absolutely convinced is wrong for an error and then finding it, when my reward is to get another manuscript in a month's time.'

Brian Conrey is fond of quoting a story told by the editor of one of the most important number theory journals, *Acta Arithmetica*, who one day received a manuscript describing a proof of the Riemann Hypothesis. He set it aside to look at when he was less busy, and then a week later he received a second manuscript from the same author, reporting a *dis*proof of the Riemann Hypothesis. In an accompanying letter the author demanded to be told which of the two was correct.

Even politics can enter the Riemann arena sometimes, as Alexander Ivic described. 'There was a mathematician in the Soviet Union…called Nikolai Gavrilov, who in the early 1960s was a big Communist boss in one of the Ukrainian cities. He was an amateur mathematician, and he got hold of the Riemann Hypothesis and thought he had a proof, and being a man of power he organized for his proof to be published. The fortunate thing is that the proof did get published. You should know that at that time the Soviets were competing, especially with the United States, in all fields – the cosmos, high technology, and so on – and they had very fine number theorists and mathematicians in general. Especially one Gelfond and one Yuri Linnik. Now, when these men got hold of a false proof of the Riemann Hypothesis, it is said, Gelfond had a heart attack and Linnik was choking and gasping for breath. He almost died from shock. They were so frustrated because much damage has been done to Soviet mathematics by this false proof of the Riemann Hypothesis appearing.'

In 1984 there was a meeting in Paris, and a Japanese mathematician, Makoto Matsumoto, gave a talk which still reverberates in the memories of several mathematicians who were working on the Riemann

Hypothesis at the time. Samuel Patterson was particularly interested in this talk because Matsumoto promised to deal with what are called metaplectic groups, mathematical objects which Patterson had worked on for a number of years.

'I had expected Matsumoto to talk in some way about metaplectic groups,' said Patterson, 'but it turned out that he essentially claimed, of a certain spherical function that he was going to produce, that if it were greater than zero then the Riemann Hypothesis was true, and he would prove that it was greater than zero. Nobody had expected this. In fact, after the talk several people went up and asked him if this was really what he was saying. And he said, "Yes." And he hadn't given any hint before the talk that this was going to be the case. The talk went on for about an hour and a half, into the lunch hour, which meant that, being in Paris, he got broken off. He had given some of the ideas but it was clear that this was by no means enough to complete it, but on the other hand he was a mathematician with a very good track record that at least I had to take pretty seriously, not someone one could regard as being a crank. I've seen the papers that he has written on this that have not been generally circulated, and they are very terse but extremely well written. In fact most people didn't take him very seriously – this is rather characteristic of mathematical life in Paris – but for me, because I had been so involved with some of his other work, I had no choice but to take him seriously.'

Alexander Ivic also heard about Matsumoto's 'proof', and was aghast: 'At the time, I was writing a book published later by John Wiley & Sons called *The Riemann Zeta Function*. It was 536 pages long. Now, anybody who writes even a crime novel of that length will understand that it's a tremendous amount of work. I submitted the book in the summer of 1984, and that fall there was a rumor that Matsumoto, a Japanese mathematician living and working in France, had produced a proof of the Riemann Hypothesis, using methods of which I knew practically nothing. So I had a friend in Princeton who kindly got hold of Matsumoto's lecture notes. He sent me the whole thing, which was even more annoying because I could not understand what was going on, except for a few first sentences or pages. So you can imagine my anguish, my frustration, at living for several months in ignorance – I had invested so much in this book. Otherwise I would be very happy and welcome the proof of the Riemann Hypothesis, definitely, but in this instance it would mean the destruction of several years of hard work. So I was very relieved when this rumor turned out to be false, and Matsumoto went to oblivion where he came from.'

In this case there was no formal recantation, just a scramble by various interested parties to 'find the mistake'. One of these was Yoichi Motohashi, who was approached by another Japanese mathematician, Kunihiko Kodaira, who was concerned about whether the proof was correct. When Motohashi investigated, it seemed that Matsumoto had made a mistake similar to Rademacher's, forty years before, in the 'proof' reported in *Time* magazine.

'Kodaida asked me to read and analyse that paper and I got the manuscript of Matsumoto's paper from him,' Motohashi told me, 'and instantly I found a mistake, because of his definition of the logarithm of the Riemann zeta function. This is rather difficult stuff – because we are taking a logarithm of complex values of the Riemann zeta function we have to be very careful, but he didn't know how to take care. And if we take his theory for granted then the Riemann Hypothesis will instantaneously come out.'

Patterson, too, spent a lot of time trying to understand Matsumoto's 'proof': 'It was plausible that he had a new idea there. I hadn't really worked on the Riemann zeta function before this. I had not considered it to be something that was really profitable as far as I was concerned, but I got involved in dealing with the Riemann zeta function as such, in order to come to terms with what Matsumoto said. I thought about it for several years until I managed to convince myself that it had to be wrong. But it was by far the most serious attempt.'

But all through the years while the claims of Matsumoto, and the Ukrainian boss, Gavrilov, and the rumours and April fool e-mails were being chewed over by the select community of top number theorists, another mathematician was quietly working away, ignored or actively shunned by his professional colleagues, on his proof of the Riemann Hypothesis, a proof to which he was putting the final touches early in 2002. He was Louis de Branges, a man to be taken seriously, one would think, because he had previously proved another famous theorem – the Bieberbach Conjecture.

7 The Bieberbach Conjecture

It would be easy to dismiss De Branges as a crank...but he has earned the right to a hearing because the early dismissals of his work on the Bieberbach Conjecture turned out to be wrong.

Joe Shipman [64]

The popular idea of mathematics is that it is largely concerned with calculations. What many people don't realize – and mathematicians at parties have given up correcting them – is that mathematicians are often no better calculators, and sometimes worse, than the average non-mathematician. An incident during my first meeting with the Franco-American mathematician, Louis de Branges, illustrated that nicely. We were discussing the idea that mathematicians did all their best work when they were young, and I asked him when he made some particular insight.

'Let's see', he said, 'It happened in 1984 and I was born in 1932. So was I over fifty? How old was I then...?'

He thought for a while, wrestling with the problem as if it was the Riemann Hypothesis itself, and then gave up (because the exact figure was unimportant, not because he couldn't do it). Even the giants of mathematics suffer from this minor disability: 'Sir Isaac Newton,' said one observer, 'though so deep in algebra and fluxions, could not readily make up a common account: and, when he was Master of the Mint, used to get somebody else to make up his accounts for him.' [65]

Ernst Kummer, another professional mathematician who lived and taught in Germany in the 1840s, was also bad at elementary arithmetic: 'One story has him standing before a blackboard, trying to compute 7 times 9. "Ah," Kummer said to his high school class, "7 times 9 is eh, uh, is uh..." "61," one of his students volunteered. "Good," said Kummer, and wrote 61 on the board. "No," said another student, "it's 69." "Come

come, gentlemen," said Kummer, "it can't be both. It must be one or the other".[66]

An American mathematician said a hundred years ago that 'Mathematics is no more the art of reckoning and computation than architecture is the art of making bricks or hewing wood, no more than painting is the art of mixing colors on a palette, no more than the science of geology is the art of breaking rocks, or the science of anatomy the art of butchering'.[67]

These anecdotes reflect the huge gap that exists between what most of us think mathematics is and the complex and abstract body of learning that is at the focus of what mathematicians today actually do. Louis de Branges's work is at the heart of that abstraction, and he has made major contributions to it.

I went to see de Branges in France, in August 2000, to spend several days discussing mathematics with him. I felt that in addition to talking to a wide range of mathematicians about their views on the Riemann Hypothesis and other aspects of mathematics, I needed to dig deeper into what makes an individual mathematician tick, and Louis de Branges struck me as someone who was at a crucial stage in his relationship with the Riemann Hypothesis.

My original plan was to go over from London for a day, to tell him about my project, and to begin to understand where he stood on the Riemann Hypothesis and what part it played in his life. De Branges was astonished that I should consider just a day, or even a day and a night, sufficient time to spend on this task.

'I was thinking that you should really spend a week here,' he said, 'then we could discuss the history of mathematics, its value to society, the state of mathematical education, things like that, as well as the Riemann Hypothesis.'

We compromised on three days, with me travelling to France on a Sunday and returning the following Tuesday night.

De Branges is a Distinguished Professor at Purdue University in Indiana, and he was spending the summer in a small village about fifty kilometres outside Paris, in an apartment that he had bought with the proceeds of a prize he won in 1989 for proving a mathematical theory called the Bieberbach Conjecture.

I could have started my researches with any of a dozen or more mathematicians who were known to be working on the Riemann Hypothesis, but I chose de Branges for the simple reason that he told me he was putting the final touches to a proof. After getting over my initial excitement, it took only two or three phone calls to other mathematicians to realize that no one else took such a statement seriously. This wasn't the first time de Branges had put the finishing touches to a proof of the Riemann Hypothesis. There was 'no chance' that de Branges had a proof, his fellow-mathematicians told me; he was 'always telling people he had a proof and it was always riddled with errors'; he's 'working in the wrong area' for a proof of the Riemann Hypothesis; and so on. I felt sure these people all believed sincerely in what they were saying, but their scepticism made me want to find out more about de Branges, as a way of beginning to understand what the Riemann Hypothesis was and what mathematicians do when they are trying to prove a major hypothesis.

Here was a man who had actually been thinking about the Riemann Hypothesis for twenty years or more. Surely he'd know it as well as any other mathematician, even if he was barking up the wrong tree in his search for a proof. And there was always the Bieberbach Conjecture to his credit.

This conjecture is a mathematical statement, first formulated by Ludwig Bieberbach, a German mathematician, in 1916. One dictionary of mathematics describes it as follows:

> The conjecture, proved by Louis de Branges in 1985, [states] that if S is the class of normalized injective holomorphic functions, so that S consists of one-to-one holomorphic functions from the unit disk with power series of the form $z + a_2 z^2 + \ldots + a_n z^n + \ldots$ for $|z| < 1$; then for every function in S, the coefficients satisfy $|a_n| \leq n$ for all n. [68]

Any non-mathematician reading this statement will find it entirely opaque. Terms such as 'normalized', 'injective', 'holomorphic', 'unit disk' and so on are not a part of normal conversation. And if, in an attempt to understand it better, the reader looks up each of the most unfamiliar terms in the same mathematical dictionary, he or she will find definitions that use the following terms in their explanations: 'orthogonal components', 'vector', 'codomain', 'set', complex derivative', 'Cauchy–Riemann', 'neighbourhood', 'metric space' and so on. Hardly more enlightening. It's clear that, just as you can't read a French book by looking up every word in a dictionary, you can't get a hold on mathematics that way either. It requires a deeper level of understanding.

'In the company of friends,' wrote psychoanalyst Alfred Adler, 'writers can discuss their books, economists the state of the economy, lawyers their latest cases, and businessmen their latest acquisitions, but mathematicians cannot discuss their mathematics at all. And the more profound their work, the less understandable it is.'[69]

Ludwig Bieberbach was an anti-Semite who believed that there was a genetic basis for mathematical thinking which resulted in a difference between German mathematicians and Jewish (and French) mathematicians. In an article explaining his views published in 1934, Bieberbach took as an example the way his two types of mathematicians thought about the theory of imaginary numbers.

'Technical virtuosity and juggling with conceptions are signs betraying the S-type, hostile to life and inorganic,' he wrote. (The S-type, according to Bieberbach, was characteristic of Jews.) In his conclusion to an article about this extraordinary theory published in *Nature*, G.H. Hardy had this to say:

> I feel disposed to add one comment only. It is not reasonable to criticize too closely the utterances, even of men of science, in times of intense political or national excitement. There are many of us, many Englishmen and many Germans, who said things during the [First World] War which we scarcely meant and are sorry to remember now. Anxiety for one's own position, dread of falling behind the rising torment of folly, determination at all costs not to be outdone, may be natural if not particularly heroic excuses. Professor Bieberbach's reputation excludes such explanations of his utterances; and I find myself driven to the more uncharitable conclusion that he really believes them true.[70]

But Bieberbach's conjecture was an important statement about functions of complex numbers, and de Branges succeeded in proving it in 1984, sixty-eight years after it was first formulated. It was no mean mathematical feat, and even mathematicians who don't take de Branges seriously today do not deny the magnitude of this achievement.

In fact, as I read about de Branges and the Bieberbach Conjecture, I began to think that the dismissive reactions generated by his work at the time were similar to what people now say about de Branges's work on the Riemann Hypothesis. As the editors of a book about de Branges's work wrote, 'In March of 1984 the message began to travel. Louis de Branges was claiming a proof of the Bieberbach Conjecture. And his method had come from totally unexpected sources: operator theory and special functions. The story seemed fantastic at the time, but it turned out to be true.'[71]

The Bieberbach Conjecture centres on a class of mathematical expressions that contains a number of terms, each involving a power of an unknown complex number, z. There could be zs and z^2s and z^3s and so on in the expression, and each term would be multiplied by a different number, a_n. So you could write the expression $z + a_2 z^2 + a_3 z^3 + a_4 z^4 + a_5 z^5 + \dots$. Bieberbach's conjecture was that these a_ns would never be larger* than the particular power of z. So z^2 would never be multiplied by a number greater than 2, z^6 would never be multiplied by a number greater than 6, and so on. (I realize that this explanation falls a little short of providing a deep understanding of the Bieberbach Conjecture, but that's all we need to know to understand the problem itself.)

Between 1916 and the 1960s, a succession of mathematicians proved the result for successively higher values. Bieberbach proved it for the coefficients of z^2, and other mathematicians for z^3, z^4, z^5 and z^6, each of them using entirely different and very complex methods. But no one could show that the general rule applied – until de Branges came along. Lars Ahlfors, of Harvard, puts the achievement in context:

> The most remarkable thing about the solution is that it was found by a known mature mathematician, not by an unknown precocious youngster. On second thought it is perhaps not so astonishing. It is absolutely certain that the conjecture could not have been proved within a few years, neither by Bieberbach nor by anybody else. For one thing, the mathematical techniques were not nearly sufficient at the time. The fact is that an enormous amount of preliminary work had to be done before any single person could have arrived at a proof. As it is, it took sixty-eight years of related research before the proof was found. Whether it could have been done in fifty years, or whether it would have taken a hundred years without de Branges, is a question which cannot be answered. But it is hardly surprising that the solution had to wait for the computer age.[72]

For de Branges, who had been working on a proof for several years, the endgame began one winter's day when he consulted a colleague, Walter

---

* In fact, to be mathematically accurate, the *absolute* value of a_n (the value if you disregard the sign, so that the absolute value of $-a$ is a) would never be larger than the power of z.

Gautschi, at Purdue University, in Lafayette, Indiana. Gautschi recalls:

> Around February 3, 1984, Louis de Branges came to my office and asked whether he could talk to me for a minute about some work he was doing; perhaps I could be of help. I distinctly remember the first thought that ran through my mind: 'Me? Helping de Branges?' We hardly knew each other, never engaged in any mathematical conversation in all the twenty or so years we were at Purdue, and – so I believed – had interests diametrically apart. He sat down and told me that he had a way of proving the Bieberbach Conjecture, but needed to establish certain inequalities involving hypergeometric functions. He felt it would be worthwhile, as a first step, to check as many of these inequalities as possible on the computer. Could I do this for him? [73]

To understand the significance of this, it isn't necessary to know what 'inequalities' (in a mathematical sense) or 'hypergeometric functions' are. What is clear is that de Branges had reached a crucial point in his work and believed he needed a final piece of information to clinch it. Gautschi was quite busy at the time, and told de Branges that he wouldn't be able to help straight away; but eventually he did. But before he used up a lot of computer time, he decided to see whether somebody had already proved the inequalities de Branges was trying to demonstrate.

'I knew there was only a handful of mathematicians in the world who could possibly be familiar with a result of this type, and even come up with a proof of it,' Gautschi wrote, 'among them Dick Askey at the University of Michigan, whom I knew best. So I called him on February 29 and told him of the inequality and what it implied according to de Branges. He immediately interrupted me with an emphatic: "I don't believe it!" and recounted some rather outrageous claims that had been made in the past by a number of people.' [74]

Askey later said that what de Branges believed 'was preposterous… Gautschi gave me the inequality and said it seemed to be true, but that it was probably very hard to prove.' In spite of his incredulity, that evening something made Askey look up a paper he had written with a colleague eight years before. 'To my surprise,' he said, 'the inequality de Branges wanted was proved in this paper.' [75]

Gautschi was working late at home on his lectures. 'The phone rang,' he said, 'and I heard Dick Askey's triumphant voice on the other end of the line: "The inequality is not a conjecture – it's a theorem!" ' *

Perhaps de Branges couldn't be blamed for not knowing that the result he needed to complete the proof of the Bieberbach Conjecture had already been proved, when one of the people who had already proved it didn't even remember himself. In fact, Askey still didn't believe that this finding would help de Branges, but Gautschi was delighted with this piece of information:

'After I checked and confirmed the result myself, I saw Louis the next morning and told him the good news. He replied, rather matter-of-factly: "Well, that proves Bieberbach's conjecture." ' And he was right.[76]

Two months later, de Branges flew to St Petersburg (at the time called Leningrad) to present his proof at the Steklov Institute, a leading centre of mathematics in the Soviet Union. He had already been invited to spend three months working at the Institute, a place where his fellow mathematicians did not seem to treat him with quite the contempt shown by some of his American colleagues.

'My stay in Leningrad,' said de Branges, 'was one of the happiest periods of my life. A typical day began with a walk through a huge park nearby. Then, after bathing, eating and shopping, I was free to do whatever I wanted. Tuesdays and Thursdays were seminar days. On those days I took the subway downtown and walked along Nevskii Prospekt until I reached the Mathematical Institute, which is located along a bank of one of the canals passing through the city.'[77]

But even here, when he came to lay out his proof before the Russian mathematicians at a seminar, he felt that they needed some convincing.

'I was introduced to Professor Milin, [and he] had an amused look which I interpreted as doubts about the soundness of my undertaking. This seemed to be confirmed a few minutes later when a question was asked about my work on the Peterson Conjecture, which contains an uncorrected error. No one seemed to believe a word of my lecture...'[78]

But then a meeting took place without de Branges, at which one or two mathematicians who knew the merits of his work convinced their colleagues.

'In a third meeting with the seminar Professor Milin shook my hand,' said de Branges, 'without losing the glint of humour in his eye, and said

---

* A conjecture is a rather tentative hypothesis, whereas a theorem is a statement that has been proved to be true.

that I was a very talented mathematician. He also said that my argument was elegant.'[79]

Among those who helped in what was the long and arduous task of verifying de Branges's proof of the Bieberbach Conjecture was Nikolai Nikolski. Today, Nikolski is a trim, wiry sixty-year-old mathematician working at the University of Bordeaux in France. He is unique among mathematicians in that he knows, respects and understands Louis de Branges. He's worked in France since 1991, having left Russia as it slipped into chaos in the Gorbachev era, but in the 1980s he was working at the Steklov Institute. Nikolski, in shirtsleeves on a hot August day, leaned back in his chair, his animated eyes gleaming through steel-rimmed spectacles. Speaking in fluent and only moderately grammatical English, he gave me a good-natured and tolerant account of what must have been a tough few years at the Steklov Institute, where de Branges's proof of the Bieberbach Conjecture was wrestled into shape.

'[De Branges] arrived in Leningrad with a manuscript of about four hundred pages containing the proof. It contained exactly a hundred theorems, and the hundredth was the proof of the Bieberbach Conjecture. The hundredth depended on ninety-nine, ninety-eight and so on, and I was in some sense the distributor of theorems to check. The people became enthusiastic. But although the proof was really there, many others did not believe in it – and me in particular – so I was very critical from the beginning.

'So during the three months a team worked away at the theorems. But they would come to me from time to time, always saying the same thing: "This theorem ninety-nine is false – here is a counter-example." De Branges would say, "No matter, this result does not depend heavily on this part." Then a week later someone else comes in, saying, "Ninety-eight is not correct – here is a counter-example," and they would give me some pieces of paper, which I kept in a drawer in my desk. In this way we got back to theorem eighty-two, and all of them were incorrect. But then one day some young person called me, and said, "Yes, I just found the ground – I found inside the theorems something that is right, so it probably could be developed into the proof," and then in two or three weeks these young people, who had not a Ph.D. between them, just very young Ph.D. students, brought the main details of the real proof.

'And when it was finished it was a great event. So the director of the Institute, Fadeyev, who became later the president of the International

Union of Mathematicians, congratulated de Branges at a big seminar. After that, I started becoming interested in how de Branges himself proved this. He became a kind of friend because during these three and a half months in Leningrad we spoke very often, walked in the city and so on. And we started to reconsider and to restate all this theory. It took two years, but we wrote two papers of eighty pages long each and so understood the real origin of this proof.'

It seems on the surface as though the proof of the Bieberbach Conjecture was more of a joint effort than a single mathematician's achievement. That's certainly how Hugh Montgomery of the University of Wisconsin sees it.

'De Branges certainly had fruitful ideas, but couldn't get them assembled without interacting with other mathematicians to critique his work. The story there is he was invited to visit Russia to give some lectures on a different topic and he said, "OK, I'll give lectures on that if you'll also allow me equal time to lecture on the Bieberbach conjecture," and they had some very good experts in Russia in this area at the time who attended his lectures and he would give a lecture and they would critique. In some circumstances in Russia, lectures can be very interactive so that the audience may interrupt the lecturer as he is talking, and so this was very interactive. De Branges would give lectures and he'd get shot down, and he'd come back the next day and give lectures and get shot down. So by the time he left Russia he had the bulk of the proof. He'd gotten it to a point where if he could prove a certain inequality then it would follow. It turned out that this inequality was already a proved inequality in the literature and so there was no need to do anything further or provide any further numerical verifications or anything, and the proof was complete.'

In fact, there's nothing unusual in a mathematician producing an initial proof that has assumptions that need to be tested. The key contribution to a major proof is the vision of the mathematician, who sees the broad outline of the proof. When the English explorer Richard Burton formed a belief in the 1850s that the source of the River Nile was a thousand miles from other suggested locations, he had a lot of work to do – and some blind alleys to explore – before his idea was proved to be correct. He also needed the help of many others to achieve his aim. But this didn't invalidate his original insight.

With plaudits ringing in his ears, from the future president of the

International Union of Mathematicians down, de Branges returned home to a situation which confirmed his jaundiced view of his mathematical colleagues in America.

It is difficult to disentangle truth from fantasy in the series of events that took place in the American mathematical community after de Branges's successful proof of the Bieberbach Conjecture. De Branges believes that, even in his moment of triumph, credit was snatched away from him by jealous and less talented colleagues.

'What happened with the Bieberbach Conjecture', he said, 'is that a major catastrophe occurred, namely that my work was not accepted. Others rewrote my proof, and they said, "This is de Branges's proof of the Bieberbach Conjecture," and then that is considered to be the proof, and my view in the matter has been rejected. Now, I've objected to this – it isn't my proof. And as I understand the history of mathematics, I don't think that anything like this has ever occurred previously. This is one of the reasons why I think that our times are different from anything in the past. I had proved a result, and I was, in my opinion, the only knowledgeable person on this subject. I had to go to St Petersburg for a verification, and brought a manuscript out, and yet despite that, others rewrote the thing and decided that was the proof.'

In the years since proving the Bieberbach Conjecture, de Branges continued to be unhappy with the reactions of his colleagues as he turned his attention to the Riemann Hypothesis. Yet, every mathematician I spoke to accepts that he is undoubtedly a good mathematician, including Peter Sarnak, of the Institute for Advanced Study.

'Bieberbach was a tremendous achievement, there's no question about it,' he said. 'Louis de Branges hit the big time there, really. It was a great problem…and his solution was absolutely brilliant, really brilliant.'

But as the details of de Branges's approach to the Riemann Hypothesis emerged, some mathematicians' high estimation of his abilities was tempered by their belief that he was wrong.

'De Branges's theory is quite nice', said Brian Conrey. 'He has a very nice theory of zeros of entire functions and things like that, it's a very beautiful theory…it's *very nice mathematics*. It just doesn't apply to the Riemann Hypothesis.'

'His work is enormously insightful and full of intelligence,' said Jonathan Keating of the University of Bristol, 'but not correct.'

I discussed his colleagues' reactions with de Branges in August 2000, as we ate sandwiches in a local bar. He is disarming in his honesty. In fact, he's even honest about how honest he is.

'I differ from other mathematicians,' he said, 'in that I seem to have a deep honesty that other people sometimes don't have, and it's rather curious because I certainly didn't have that intention as a young man. I think I was by nature somebody that would easily cut a corner – especially if I didn't think it was very important – not for any real advantage, but I would choose to do so. But the way my life has evolved, against my own inclinations, I have turned out to have an unusual probity.'

It would be easy to dismiss such remarks as those of someone who keeps telling you how modest he is, with the aim of proving that he's the most modest man alive. But there were often times when he told me things that other people would think twice about before revealing.

'My mind is not very flexible,' he told me. 'I concentrate on one thing and I am incapable of keeping an overall picture. So when I focus on the one thing, I actually forget about the rest of it, and so then I see that at some later time the memory does put it together and there's been an omission. So when that happens then I have to be very careful with myself that I don't fall into some sort of a depression or something like that. You expect that something's going to happen and a major change has taken place, and what you have to realize at that point is that you are vulnerable and that you have to give yourself time to wait until the truth comes out.'

The time we first met, de Branges was a little preoccupied. He had been having lessons over the last few weeks, two or three times a week, in the French Highway Code. In order to be allowed to drive in France he had to pass an exam, because his Indiana driver's licence was not recognized. I was surprised to discover that he was nervous about this exam, and even more surprised when he told me about some of the questions that were giving him trouble. There was a question on speed limits, which showed a sign with '80' on it, and asked which of three speeds you are permitted to travel at after passing the sign: 60, 80, 100 k.p.h. During his lessons, in which he was set test questions, de Branges kept making the mistake of answering '80' when he should have answered '60 and 80' (because clearly if you are permitted to travel at 80 k.p.h you are also permitted to travel at 60). Perhaps it isn't surprising that you should make such a mistake the first time you are asked, but de Branges was having difficulty remembering that he had to answer with both speeds each time he was asked this question. 'I keep making errors,' he said.

After the unhappy events in America that culminated in a symposium about his proof where he felt ignored, de Branges made a decision that was to dominate the next fifteen years of his working life. He decided to start working immediately on the Riemann Hypothesis.

'I had been inspired in St Petersburg. It turns out that there is some sort of deep relationship between the Bieberbach Conjecture and the Riemann Hypothesis that's not completely understood, so I think the final truth will be that there will be a unity with the Riemann Hypothesis. But that unity cannot for the moment be described and materialized. And I saw that at this point in my career, when I was, I think over fifty at that time, that I couldn't afford to let this go; that I would have to put the rest of my life entirely, on this thing.'

As I spoke to de Branges and others about his work on the Riemann Hypothesis and why it is universally dismissed, I kept coming across parallels with earlier reactions to his Bieberbach Conjecture work. When Richard Askey looked back at why he was initially sceptical about what he called de Branges's 'marvellous proof', he said a number of things that struck me as a very accurate foreshadowing of the kind of reactions de Branges was to hear and read as he reported his work on the Riemann Hypothesis:

> There is a lot to be said for doing two things de Branges did. He worked on a hard problem, has worked on other hard problems in the past, and was not discouraged by past failures. Also, it often pays not to talk with experts in an area, or at least not to take their pessimism too seriously. Experts in univalent functions would have told de Branges that his method could not work, and they would have been wrong. My reaction was similar. I knew too much in one direction, where my optimism had led to a number of deep inequalities, and I did not know enough in another area. These two led to a skepticism that was unwarranted. Do not take the pessimistic views of an expert very seriously, but optimistic views should be taken very seriously, for they often lead to something important... The important fact is de Branges's proof, not whether he had published false proofs of other conjectures. The fact that de Branges was an outsider in univalent functions is important, but no one who knows his early work can say he

was on the fringe of the research community. If one does not know, then it is best to say nothing. [80]

Much of this assessment could be applied directly to de Branges's work today.

'*He worked on a hard problem, has worked on other hard problems in the past, and was not discouraged by past failures.*' All these factors were to blight de Branges's attempts to get his efforts on the Riemann Hypothesis recognized.

'*… it often pays…not to take [experts'] pessimism too seriously. Experts in univalent functions would have told de Branges that his method could not work, and they would have been wrong.*' When experts in number theory bothered to communicate with de Branges at all, they usually told him that his approach to the Riemann Hypothesis would not work.

'*The important fact is de Branges's proof, not whether he had published false proofs of other conjectures.*' The erroneous proofs he had published, both of the Bieberbach Conjecture and the Riemann Hypothesis, were to colour overwhelmingly the attitude of his peers and colleagues to his later publications.

In fact, Atle Selberg, a veteran number theorist, showed what harm de Branges's mistakes have done to his image. 'The thing is, it's very dangerous to have a fixed idea. A person with a fixed idea will always find some way of convincing himself in the end that he is right. Louis de Branges has committed a lot of mistakes in his life. Mathematically he is not the most reliable source in that sense. As I once said to someone – it's a somewhat malicious jest but occasionally I engage in that – after finally they had verified that he had made this result on the Bieberbach Conjecture, I said that Louis de Branges has made all kinds of mistakes, and this time he has made the mistake of being right!'

It has to be said, too, that that many people see de Branges as brusque, preoccupied, obsessive and stubborn. All in all, it's not difficult to see how it is that many colleagues might be deterred from giving his recent work on the Riemann Hypothesis the attention he believes it deserves. The question at the beginning of the twenty-first century was this: is de Branges about to pull off the same trick with the Riemann Hypothesis as he did with the Bieberbach Conjecture, or are most of the world's number theorists right in dismissing his latest efforts?

8 In search of zeros

Numbers are friends to me, more or less. It doesn't mean the same
for you, does it, 3,844? For you it's just a three and an eight and a
four and a four. But I say, 'Hi, 62 squared!'

Wim Klein [81]

The interior life of mathematicians is a closed world to the rest of us.
How can a non-mathematician imagine sitting down to read 'cover to
cover', as one fan has done, a book called *A Handbook of Integer
Sequences*? Published in 1973, the book was so popular that a successor
volume was produced twenty-two years later, called *The Encyclopedia
of Integer Sequences*.[82] The books contain pages and pages of numbers,
arranged so that if you come across an unfamiliar sequence of numbers
in your daily life you can look it up and find out if there is a rule that
generates it. (Its Chapter 2, in the manner of a survival guide for explorers, is called 'How to Handle a Strange Sequence'.)

Let's open the book and see where its charm lies. The very first
sequence of numbers is $0, 0, 0, 0, 0, 0, 0, 0, 0, 0, 0, 0, 0, 0, 0, 0, 0, 0, 0, 0, \ldots$.
This, in case you wondered, is 'the zero sequence'. Skipping over the
next two sequences, which are less interesting, we come to $1, 1, 1, 1, 1, 1,$
$1, 1, 1, 1, 1, 1, 1, 1, 1, 1, 1, 1, 1, 1, 1, 1, 1, 1, 1, 1, 1, 1, 1, 1, 1, 1, \ldots$.
This, apparently, is 'the simplest sequence of positive numbers'. The
excitement increases, however, a few items later when we find $1, 1, 0, 1,$
$2, 0, 0, 1, 1, 2, 0, 0, 2, 0, 0, 1, 2, 1, 0, 2, 0, 0, 0, 0, 3, 2, 0, 0, 2, 0, 0, 1, 0, \ldots$.
This turns out to be the number of ways of 'writing n as a sum of less
than or equal to two squares'. Just to reinforce the idea that there might
be something not entirely serious about the exercise, there's a series that
begins $1, 3, 7, 19, 53, 149, 419, \ldots$, which is the number of stable towers
that can be built from n Lego blocks; while $1, 2, 4, 7, 11, 16, 22, \ldots$ is the
maximum number of pieces you can get from a pancake by slicing it
with n cuts.

Bearing in mind that primes play an important part in this book, you might like to guess what this sequence represents:

2, 11, 101, 1,009, 10,007, 100,003, 1,000,003, 100,000,019, 100,000,007, 1,000,000,007, 100,000,000,019, 100,000,000,003, ...*

Some sequences in the *Encyclopedia* are rather inadequately annotated. The five numbers 8, 28, 89, 234, 512 apparently constitute the 'postage stamp problem'. A sequence beginning 2, 10, 74, 518, sounds rather charming: it is '$(2n+1)$-step walks on a diamond lattice'. For the Satanists among its readers, the editors of the *Encyclopedia* include 157, 192, 218, 220, 222, 224, 226, 243, 245, ... This sequence of numbers contains all the powers of 2 that have within them the number 666. (So somewhere in 2^{222}, if you can be bothered to work it out, are the digits '666'.)

The *Encyclopedia* is a tribute to the immense versatility and complexity of our number system. Many of the sequences play a fundamental part in number theory and related fields, and people who spend their lives immersed in numbers can perhaps be forgiven for the occasional whimsical diversion. What the book shows, in thousands of entries I have not quoted, is that in a sense the real numbers provide fertile ground for explorers. There is even a movement by some mathematicians towards what is called 'experimental mathematics', a field in which discoveries are made – usually with computers – by digging around in sequences of numbers, looking for patterns and testing hypotheses.

One mathematician who is often seen as an experimental rather than a theoretical mathematician is Andrew Odlyzko, an American originally from Poland, who worked until 2001 for AT&T's Research Laboratories in New Jersey. On a mild June afternoon in 2000, Odlyzko sat at his computer screen and showed me a list of numbers, thirty or so, filling the screen. They had been calculated by a powerful set of linked computers humming away in the basement of his office building. Each number was calculated to twelve decimal places, and he was showing me a tiny sub-sample of a much larger list. The full list contained 15,000,000,000,000 numbers, and as I listened to Odlyzko talk, and looked out over the landscape that surrounded his office, the computers were assiduously calculating another billion or so more of these numbers.

* Each number in this sequence is the smallest *n*-digit prime.

Odlyzko is tall and nearly bald, with a domed forehead and rimless spectacles. He can seem rather solemn, unworldly even, but as I entered his office I heard him arguing on the phone with a local airline about his frequent flier mileage, which brought him down to earth a little. But mathematicians do suffer from an image problem. The people who believe (wrongly) that mathematicians are a bit like accountants assume that they do sums all day. Those of us who know a bit more about what they really do are amazed at how abstract their mental lives can get, and we wonder how they can fit the mundane business of everyday life into such a rarefied existence.

At AT&T's Research Labs Odlyzko worked on the problems of telecommunications systems, but he is really a number theorist, someone who thinks all day about the properties of the system of numbers we use. It's difficult to convey to a non-mathematician how real an existence these numbers have to a mathematician. I watched Odlyzko giving a lecture on the prime numbers. For an hour he spoke off the cuff, showing a succession of overhead slides with calculations, quotations and even cartoons, and it was as though he was describing the exploration and taming of a strange landscape. Odlyzko is one of the diggers of mathematics, though it's often the guys who wave a hand airily over a specific tract of land who get the credit and the prizes when the treasure is unearthed. If Odlyzko ever finds treasure, the discovery is going to disappoint many of his colleagues. Like Alan Turing, but with infinitely greater means at his disposal, Odlyzko could one day prove that the Riemann Hypothesis is wrong – but never that it is right.

All Odlyzko's attention is focused on the zeros of the Riemann Hypothesis. He is compiling an ever-growing list of the values of s that make $\zeta(s)$ equal zero. He could prove that the Riemann Hypothesis is wrong if he discovered zeros that were not on the critical line.

For some mathematicians, such as Steve Gonek, this would be a disaster. 'If,' he said, 'there *are* lots of zeros off the line – and there might be – the whole picture is just horrible, horrible, very ugly. It's an Occam's razor sort of thing, you either have absolutely beautiful behaviour of prime numbers, they behave just like you want them to behave, or else it's *really* bad.'

Of course, it's the prime numbers that people really want to understand. But the Riemann Hypothesis, if true, promises such an intimate connection between the zeros and the primes that most people feel we'll never understand one without understanding the other. But the connection is not obvious. There is not, for example, one zero for every prime. Samuel Patterson compares the relationship between zeros and primes to the relationship between the sound of a full orchestra and the separate sounds of the instruments.

'There's a technique in mathematics called Fourier transformation,' he explains. 'The most obvious example of this is in acoustics, where you hear a particular change in pressure as a sound that you can resolve into its different frequencies. This happens in music – when you're playing a musical instrument, it's actually synthesizing sounds through the different frequencies.'

Joseph Fourier was born in Auxerre in France in 1768, and lived through and supported the French Revolution, being an active member of his local Revolutionary Committee. Later he became a scientific adviser to Napoleon on his expedition to Egypt. Fourier devised a method by which any complex waveform can be represented as the sum of a series of much simpler waves. The simplest and most symmetrical wave, from which all others can be formed, is what's called a sine curve, as shown in Figure 11.

There can be many different sine curves, depending on the distance between the peaks (or troughs). The measure used is 'frequency', and the higher the frequency of such a curve the closer together are the peaks. Fourier proved that any complicated curve can be broken down into a series of sine curves. So a curve like the one shown at the top of Figure 12 can be obtained by 'adding together' a series of sine curves of different wavelengths – the three simple curves below it. Taking the values of all three simple curves (i.e. the height or depth of the peak or trough) at corresponding points and adding them together gives you the value of the corresponding point on the complicated curve.

FIGURE 11 A simple wave – a sine curve, the building block of more complex waves in physics.

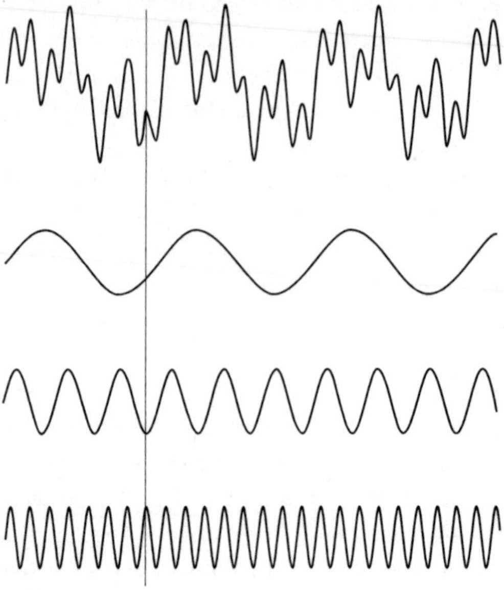

FIGURE 12 Every point in the complex curve at the top is obtained by adding together the values of corresponding points on the three sine curves below.

Because complex waveforms feature in many different branches of physics, including sound, light and radio, Fourier's technique for resolving a waveform into its components is one of the most important discoveries in mathematical physics.

'There is a historian of mathematics', said Samuel Patterson, 'called Erwin Neuenschwander, who goes around asking people who is the most named mathematician in the world, and if you look at it, it's easily Fourier. Basically the zeros of the Riemann Zeta function are giving the Fourier transform of the set of prime numbers.'

As we've seen, the distribution of prime numbers generally follows Gauss's prediction, but if you look at the distribution in detail you find that the number of primes you would expect there to be in a certain interval is usually more or fewer than the Gauss figure. Those fluctuations can be represented as a meandering wave, and that meandering wave, if subjected to a type of Fourier analysis, can be decomposed into a series of waves each of which is related to a zero of the Riemann zeta function.

The curve of primes is like the sound of the orchestra, and the zeros are like separating that sound into individual frequencies, so that each zero contributes to all the primes or each prime contributes to all the zeros. If you were to subtract the individual pure sounds from a recording of an orchestra one by one, the piece of music would still be there, sounding thinner and thinner but still recognizable, until the last sound was removed.

There's another analogy that helps. If you hold up a hologram and look 'through' it, you will see a three-dimensional object or scene beyond it. It is as if you were looking through a hole the size of the hologram. Now, if you break or cut the hologram in half and look through it, you won't see half the scene. You'll see the whole scene, laid out in the same way as before, but you'll be looking through a smaller 'hole'. Cut the hologram in half again, and you'll still see the same scene. This may seem surprising until you realize that it's a bit like looking through a window in a wall. If half the window is blocked off, you can still see all of what's on the other side.

Well, the entire collection of Riemann zeros is like a hologram of the primes. If we were able to 'look through' all the zeros we'd see an exact image of the primes, and could discover anything we wanted to know about them. But because we have only some zeros, and can never have an infinity of them, we get a picture of the primes which is not sharp enough to answer all our questions. And, who knows? If some of those zeros turn out not to be where we think they are – on the critical line – as the picture of the primes gets sharper, we might find that the details of the primes are not at all what we expected. This discovery could come from the work of Andrew Odlyzko.

Peter Sarnak at Princeton quaintly describes Odlyzko's computers as machines. 'If this thing [the Riemann Hypothesis] were false you could have a machine disprove this at any moment. You have to believe it's false to have your machine run all day and all night. Odlyzko is looking at the 10^{20}-th zero, and maybe if he goes a little higher his machine could come out with a statement that proves that the Riemann Hypothesis is false. If there were a zero off the line it could actually dispel this hypothesis. This would of course be a disaster, but it could happen.'

Because of the complexity of the zeta function and the difficulties of doing large calculations with complex numbers, Odlyzko has to use a couple of tricks to find the positions of the zeros that are on the critical line and to see whether there are any that are not.

To understand what Odlyzko's computers are doing, it's worth recapping on what it actually means to talk about the *zeros* of the Riemann zeta function. The zeta function itself – the sum of terms of the form $1/n^s$ as n goes through 1, 2, 3, and so on – has different values depending on which complex number you substitute for s. Like Charles Ryavec's Yellowstone geyser, if you throw in certain ss the zeta function throws out a zero, and if you throw in other ss the zeta function gives non-zero values, some positive, some negative.

Odlyzko uses a transformed version of the zeta function. It is possible to prove that if there are any zeros off the line, they are placed symmetrically either side of it. With Odlyzko's transformed graph, instead of the Riemann zeros being symmetrical about the line (or 'street') equal to $\frac{1}{2}$, they centre on the zero axis, producing a graph that represents the values of the zeta function as a meandering line. Each point where this line crosses the zero axis is a point where the value of the zeta function equals zero.

Since we're interested only in the values of s at the points where the curve crosses the line – where the function equals zero – we don't need to find a way to draw the whole curve. This would require us to calculate the value of the function for every possible value of s. Since we could substitute any of an infinite number of values for s, this would be impossible. But there is a short cut, a method that dates back a hundred years and might even have been used by Riemann himself. There's a way to find out how many times the graph goes from positive to negative or vice versa. Each time this happens, the curve will cross the zero line, and so the value of the function will be zero.

Odlyzko devised a method based on Riemann's original calculations as discovered by Siegel in 1932, but much faster. He first set a Cray computer the task of finding 100,000 zeros using the original version of the Riemann–Siegel method. This took 15 hours. Then, using the new method and starting with a larger value of s, the same computer found 16,000,000 zeros in 18 hours – 160 times as many zeros for just a 20 per cent increase in computer time. Their method worked by calculating the number of changes in sign of the zeta function in regularly spaced intervals on the number line. By counting these sign changes they could be pretty sure that the number of sign changes matched the number of zeros on the line.

This is fine as far as it goes. It certainly enabled Odlyzko to amass a larger collection of zeros of the Riemann zeta function than anyone else in the world had managed. And it confirmed that many millions of zeros are indeed on the critical line. But this method alone doesn't prove that there aren't zeros *off* the line – a situation that would disprove the Riemann Hypothesis at one blow. To do that, there is another technique. Odlyzko used a method which counted all the zeros in a 'box' around the interval. If there were any zeros not on the line, like the dots above and below the line in Figure 13 (such zeros, if they exist, would come in pairs either side of the line), then the box method would yield a different total from the 'sign change' method. But if both methods gave the same answer, this would prove that all the zeros in that interval were on the critical line, that there was no zero lurking somewhere off it.

As Odlyzko sits at his office computer working on some other topic, there is an alarm system that will alert him if ever the system discovers a discrepancy. 'Very often the program flags a particular region, and says

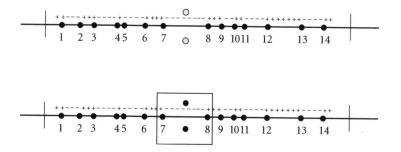

FIGURE 13 One method Odlyzko and others use to find Riemann zeros is based on the number of times the value of the Riemann zeta function changes sign from plus to minus (solid dots 1–14 above). But this method only detects zeros with real part equal to half ($\frac{1}{2} + ib$). There may be zeros with real part equal to some other number and so the sign change method wouldn't detect zeros that are off the critical line (grey dots below), the very zeros that would disprove the Riemann Hypothesis if they existed. So a back-up method is used, which counts zeros in a 'box', on or off the line. If there's a discrepancy between the two counts, the Riemann Hypothesis is false.

there is a possible violation of the Riemann Hypothesis here. That is because the program is fairly stupid and there are so many pathologies among zeros, I have to go in by hand and see what happens. It doesn't happen too often, every few million zeros, every few dozens of millions of zeros, so I have not actually gone through all those zeros. I think I have done it for about 5 or 6 billion zeros, and the remainder still remain to be fully verified.'

In fact, since I first met him, Odlyzko has worked his way higher up the list and has now dealt with all the 20 billion zeros near the 10^{23}-rd zero, looking for a counter-example. Worryingly – for believers in the truth of the Riemann Hypothesis – in the summer of 2002 there was a hint that a counter-example just *might* be lurking nearby.

It's easy to look upon Odlyzko's pursuit as train-spotting – gathering odd specimens from the number field, judging the merit of his collection by the number of items it contains. For example, he wrote a paper entitled 'The 10^{20}-th zero of the Riemann zeta function and 175 million of its neighbors'. This was clearly such a success that he followed it up with a sequel called 'The 10^{22}-nd zero of the Riemann zeta function'. (If you're interested, the 10^{22}-nd zero begins 1,370,919,909,931,308, 226.490 240... .) But for Peter Sarnak, Odlyzko contributes much more to the cause of the Riemann Hypothesis than just numbers.

'He's a great theoretician,' said Sarnak. 'He's not a hacker, he's got a lot of great theorems to his name. It's just he's unusual in that there are not many mathematicians who work in this way with computers. If Gauss were alive today I would say he'd be at the computer all the time. In other words, he wouldn't be proving things, he would be telling you what's true and experimenting and saying "OK, I now understand this" and moving on to the next thing, which is what Odlyzko's willing to do. The computer has a big role to play. If you have a hunch, you experiment with a pattern and then the theory gets to be built around those things and big problems sit there to be solved. I believe in pure math having a laboratory side, and counting primes is an example.'

There is a growing role in theoretical mathematics for computers as more than just number-crunchers. Back in 1977, the Four-Colour Theorem, which had been puzzling mathematicians for over a hundred years, was finally proved – by a computer. The theorem states that any two-dimensional map can be coloured using only four colours in such a way that regions sharing a common boundary (other than a single point) do not share the same colour. Generations of mathematicians tried to prove it, and a few thought they had succeeded. One of them, Alfred Kempe, published a proof in 1879 which was shown to be erroneous eleven years later. But a variant of the method underlying Kempe's proof was used by Kenneth Appel and Wolfgang Haken in 1976, who worked out that every map on a plane could be classified as one of 1,500 different types. They then programmed a computer to show that for each one of those types the Four-Colour Theorem was true. It took 1,200 hours of computer time to compile the final proof, and that – for many mathematicians – was a sticking point in accepting it as a proof at all: it is impossible for a human mathematician to check every detail of the computer proof of the Four-Colour Theorem. Now, over two decades later, the first ever computer proof of a major theorem has entered the annals of mathematics and is generally accepted. But it hasn't stopped mathematicians from looking for a simpler proof that humans can work through for themselves.

The fact that a computer can be used to generate a proof – in very particular circumstances – shouldn't be confused with the regular use of computers to *confirm* hypotheses rather than prove them. This is a generally accepted procedure that at least gives people the confidence to pursue a proof, and sometimes stops them in their tracks by providing a counter-example. However large the number of confirming examples, there's no guarantee that there isn't a surprise around the corner.

There's an extraordinary story about the distribution of prime numbers that illustrates this well. Earlier in the book we met Gauss's suggestion that the number of primes less than any number n (the symbol $\pi(n)$ is used to represent this number) is approximately $n/\log n$. There's a refinement of this approximation which uses a different sort of logarithm, called an integral logarithm, denoted by 'Li'. Gauss discovered that $\pi(n)$ is even closer to Li(n) than to $n/\log n$. When mathematicians used this method to calculate values of $\pi(n)$ for higher and higher values of n, they found that whereas the results were by far the best ever obtained, the difference between $\pi(n)$ and Li(n) got larger and larger as n increased. When n is 10^5, Li(n) is too large by 38; when n is 10^6, Li(n) is too large by 130; when n is 10^9, Li(n) is too large by 1,701; and so on. This suggested that Li(n) was always a little larger than $\pi(n)$, but in 1914 Littlewood proved that once some value of n had been reached, the balance switched, and Li(n) became *smaller* than $\pi(n)$ as n

increased further. Littlewood's proof did not provide any way of finding a value for n at which the switch takes place. Then, in 1933, a mathematician named Stanley Skewes found what is called an 'upper bound' for the value of n at which $\mathrm{Li}(n)$ became smaller than $\pi(n)$. This means that he proved that the switch from larger to smaller happened for some n smaller than the number he found. Skewes's number was the following:

$$10^{10^{10^{34}}}$$

Malcolm Lines, an author of books on popular mathematics writes:

> The sheer enormity of this number is difficult to appreciate. It has been called the largest number which has ever served a useful purpose in mathematics, and mathematical numbers always exceed by far those used in the other sciences. For example, the total number of atoms in the entire universe is 'only' about 10^{75}. This last number written out in full is a 1 followed by 75 zeros. The monstrous number above on the other hand has no less than $(10^{10})^{34}$ zeros. This means that, even if we used as many zeros as there are atoms in the universe, we still could not come close to writing it out in full. This seems to bode ill for anyone seeking to find the actual value of n for which the number of primes first becomes larger than its theoretical estimates. Be this as it may, the example does show us how unwise it can be to jump to conclusions on the basis of numerical data out to a puny few billion.[83]

Hardy used Skewes's number as an example in which 'the truth has defeated not only all the evidence of the facts and of common sense but even a mathematical imagination as powerful as that of Gauss', and he tried to convey how large the number is, in human terms. 'The number of protons in the universe is about 10^{80},' he wrote. 'The number of possible games of chess is much larger, perhaps $(10^{10})^{50}$. If the universe were the chessboard, the protons the chessmen, and any interchange in the position of two protons a move, then the number of possible games would be something like the Skewes number.'[84]

This number is an upper bound, remember, meaning that the exact point at which $\mathrm{Li}(n)$ becomes smaller than $\pi(n)$ is no larger than Skewes's number. But it could be smaller, and over the years other mathematicians tried to get an upper bound that was lower than Skewes's original estimate. In the late 1960s, R. Sherman Lehman at the

University of Berkeley used a different method to reduce the upper bound to about $10^{1,130}$. That number was reduced in the 1990s by Herman te Riele to something like 10^{370} or 10^{380}, and then more recently it was reduced by Hudson and Bayes to something like 10^{310}. At the same time, working from the bottom upwards, mathematicians have calculated that there is no counter-example below 10^{12}. So the exact answer to the problem is somewhere between 10^{12} and 10^{310}, one number with 12 zeros and another with 310.

Pondering the mystery of Skewes's original upper bound, Littlewood asked whether there was some answer to a mathematical problem which could be proved to exist but which was too large 'to be mentioned', as he put it. What Littlewood meant by 'mentioned' was the possibility that there might be a number so large that it was impossible to write it down or indicate its magnitude in any conventional way. It turns out that there is such a number, discovered by the mathematician Ronald J. Graham. It's another upper bound, obtained when Graham was trying to solve a problem posed by the English mathematician Frank Ramsey (who died tragically young in 1930, at the age of twenty-six, having already achieved major results in mathematics, philosophy and economics).

One of Ramsey's many enduring achievements was a branch of mathematical logic called Ramsey theory. It dealt with questions such as finding the minimum number of people for there to be at least two of one sex or the other. That one's not too difficult. Another less obvious statement that can be proved with Ramsey theory is that in any collection of six people, either three know each other or three of them do not know each other. The Ramsey problem that led to Ronald Graham's large number was this:

> Take any number of people, list every possible committee that can be formed from them, and consider every possible pair of committees. How many people must be in the original group so that no matter how the assignments are made, there will be four committees in which all the pairs fall in the same group, and all the people belong to an even number of committees?

Ronald Graham found an upper bound – let's call it U – for the answer to this problem. He proved that the number of people there must be in the original group to fulfil Ramsey's conditions could not be larger than

U. But U was so large that it couldn't be written in normal mathematical notation. Suffice it to say that if I said that Skewes's number is to Graham's number as an atom is to the entire universe, I would be getting nowhere near the truth. And to cap the whole story, most experts in Ramsey theory today believe that the correct answer is 6.

Odlyzko's work on the Riemann zeros raises a similar issue of large numbers. As he probes to higher and higher values of zeros, do the chances get any greater that he will come across a zero that is off the line? Alexander Ivic, a Serbian mathematician who has several significant number theory proofs to his credit, believes that such a zero might exist but that there is not and never will be enough computing power in the world to find it.

He is a blunt-spoken man who rolls his r s and declaims as he speaks about the subject dear to his heart. 'Now, as much as this is a possibility, the overwhelming evidence in favour of the Riemann Hypothesis – namely that one and a half billion zeros satisfy the Riemann hypothesis plus many more zeros of greater height – has one drawback. Numerical calculations cannot go too far in the present state of computers. All you can hope for with the present state of computers if it goes on for twenty, thirty, maybe fifty years more is to take into account a zero of height 10^{40}. Now that's a huge number by all means for us mortals, but not when you take into account that in formulas involving the zeta function we have the iterated logarithm.'

The 'iterated logarithm' is the logarithm of a logarithm, and the iterated logarithm of x grows very slowly as x increases. So if you consider 10^{40}, say, the log of this number to the base 10 is 40, and the log of the log is about 1.6. This means that as the numbers get bigger and bigger, the log of the log varies much more slowly. Go from 10^{40} to $10^{1,000}$ – a huge increase – and the log of the log increases from 1.6 to 3. Ivic went on to describe how the region where such 'off-the-line' zeros might occur was in the farthest reaches of the line of zeros, determined by the slow growth of the iterated logarithm.

'So for all practical purposes, the Riemann zeta function does not show its true colours in the range available by numerical investigations. You should go up to the height $10^{10,000}$, then I would be much more convinced if things were still pointing strongly in the direction of the Riemann Hypothesis. So numerical calculations are certainly very impressive, and they are a triumph of computers and numerical analysis, but they are of limited capacity. The Riemann Hypothesis is a very delicate mechanism. It works as far as we know for all existing zeros, but we cannot, of course, verify numerically an infinity of zeros, so other theoretical ways of approach must be found, and for the time being they are insufficient to yield any positive conclusion.'

Odlyzko is neutral about the truth of the Riemann Hypothesis. 'Quite a few people don't want to believe it's true. Littlewood is thought to have died disbelieving it. Personally I am also not absolutely sure that it's true. [But] I'm fairly confident that if there is a counter-example it's way beyond the reach of what I'm doing.'

So if Odlyzko is not really working on a proof or disproof of the Riemann Hypothesis what is the benefit of his detailed analysis of the zeros? Why do so many mathematicians value his work so highly? Michael Berry for one.

'He started doing these wonderfully epic computations,' Berry said enthusiastically, 'and he got beautiful results.' But Berry's excitement comes from a very specific series of discoveries that have emerged from Odlyzko's calculations, discoveries which, if they don't themselves prove the Riemann Hypothesis, may well put the means of proof into the hands of Berry and his colleagues. It's the latest episode in a story that began with a tea party in Princeton in 1972.

The Princeton tea party

That's the density of the pair correlation of eigenvalues of random matrices in the Gaussian Unitary Ensemble...

Freeman Dyson to Hugh Montgomery

Mathematics – and indeed science in general – is often seen as a collaborative enterprise on a global scale. Most advances come about through the laborious assembly of many separate pieces of knowledge, and rarely have those pieces of knowledge all come out of the same brain. On the whole, there is more benefit to an individual mathematician from sharing results than from concealing them. So when Hugh Montgomery was visiting Princeton in 1972 and was introduced to the polymath scientist and mathematician Freeman Dyson over tea, he answered perfectly truthfully when Dyson asked him conversationally what he was working on. His answer struck a chord with Dyson, who then supplied a piece of information that indirectly led to what today is seen by many as the most promising approach to proving the Riemann Hypothesis.

The Institute for Advanced Study on the Princeton campus is where Einstein spent his closing years, after he had become the most famous scientist in the world. In fact, there is a book about the IAS called *Who Got Einstein's Office?*, a question often asked by visitors. The man who *did* get Einstein's office, a large room with tall bay windows looking out over grassy lawns, is rather fed up with the attraction it generates as Japanese tourists snap away outside.

In the maths department of the Institute are to be found the purest examples of unfettered mathematizing. Off its corridors are pleasant, light offices, each with a blackboard, a desk and a mathematician. The mathematicians are doing one of three things – staring out of the window, writing furiously with a pencil or scribbling with chalk on a blackboard. This is what they are paid to do, and they appear to be doing it very well. This is, I suppose, what also goes on in a literature or history department at any university. But there is one difference – on the whole, historians, literary critics and even writers are recycling material; adding their own spin, certainly, but essentially recycling facts, words, emotions, descriptions that usually exist in some form already. Mathematicians, however, are innovators. There is little pleasure or profit to be gained in reconsidering proofs of theorems from the past, and not much more in finding new ways of proving the same theorems. But discovering new relationships between numbers, exploring new or partly analysed functions, solving knotty and frustrating problems – all these can lead to concepts that no mind has ever encountered before.

Apart from Einstein there are very few famous modern mathematicians, people whose achievements are known to the public. One of the few is Andrew Wiles, now working at the IAS, who in 1995 at the age of forty-four proved Fermat's Last Theorem. This theorem was first posed by the seventeenth-century French mathematician, Pierre de Fermat, who claimed that he had a proof but never published it.* The story of Wiles's proof generated extensive media coverage including a television documentary and at least two popular books. If the Riemann Hypothesis is ever proved, it is likely to attract considerably more.

In November 2000, as Peter Sarnak and I walked across the IAS lawn to the cafeteria for lunch, we ran into Wiles (now Sir Andrew), a quietly spoken, diffident and secretive man for whom fame seems not the spur but a rein on his future work. You could not imagine a man less willing or able to rise to the demands of hungry media and fascinated – but uncomprehending – non-mathematicians.

Peter Sarnak was one of the few mathematicians whom Wiles kept informed during the later stages of his attack on Fermat's Last Theorem.

'The very first thing he told me when he came to my house,' Sarnak said, 'was "I think I've proved Fermat's Last Theorem", and I said, "Oh, yes, I think I've proved Riemann." I was joking with him. And then he just said, "Well, let me show you an idea." And after ten minutes I couldn't move, it was clear that this was a *major idea*, and that this was a major breakthrough even if it didn't work. I haven't seen that with the Riemann Hypothesis. We'll know when something like that comes

---

* Fermat stated without proof that there was no larger power than $n = 2$ that would make the expression $x^n + y^n = z^n$ true for whole numbers x, y and z.

along because it will be clear that the guy's on to some fundamentally new idea.'

As we all had lunch together, Sarnak and Wiles were discussing a musical that had recently opened in Manhattan.

'It's amazing how they got Shimura–Taniyama into one of the songs,' Sarnak said.

'I haven't seen it,' said Wiles. 'Apparently they have a woman playing my wife although they've never met her.' It seemed unlikely, but this musical, called *Fermat's Last Tango*, had been written by a composer and a writer who had seen the TV documentary, read the book and believed that Wiles's story was the stuff of drama.

That evening I sat in a theatre and watched open-mouthed as a cast of seven sang words that I never dreamed I would hear on a stage in front of a non-specialist audience. The audience was a mix of mathematical ability – from zero to advanced, to judge by reactions to some of the songs and jokes. The Wiles character sings lines such as

> I knew, I swore,
> That elegant symmetry
> Of x squared plus y squared
> Is square of z
> Could not be repeated if n were three,
> Or more!

At some point, Fermat appears and pours scorn on Wiles's proof, using mathematical tools that were not available to the seventeenth-century mathematician, singing

> Elliptical curves, modular forms,
> Shimura–Taniyama,
> It's all made up, it doesn't exist,
> Algebraic melodrama!

In the steps that led up to Wiles's proof of Fermat's Last Theorem, there was a mistake that Wiles didn't spot until after he had announced his proof. In the musical, Fermat points out Wiles's mistake in perhaps the most impenetrable lines of script ever spoken by an actor in front of a general audience:

> In order to transform your elliptic curves into Galois representations so
> they could be counted against the set of modular forms, you assumed they
> met the requirements of an Euler system, when in fact they do not!

The most uncanny part of the musical was the performance of the actor
who played Wiles. Resisting the temptation to create a conventional
heroic character in place of the distinctly low-key, real-life mathematician, the producers had gone for an accurate portrayal of Wiles, under
the name of Daniel Keane, and this made his mathematical achievement
all the greater. Comfortable clothes, diffident gait, low-key delivery,
absent-minded concentration on his work – the actor seemed remarkably like the man I had met at lunch that day. Portraying your hero as a
nerd could backfire and turn him into a joke. Here, somehow, the Wiles
character was elevated by his achievements.

The writers of this musical had managed to capture the essence of the
mathematical enterprise and to see that the human drama of Wiles's
struggle with Fermat's Last Theorem embodied as much passion, frustration and triumph as is found in the plot of any conventional film or
play. It turns out that if this is true of Fermat's Last Theorem, it's likely
to be true – in spades – of the Riemann Hypothesis. Take the issue of
secrecy, for example. For all the magnitude of Wiles's achievement, there
were some mathematicians who felt that their own professional lives
had been blighted as a result of him playing his cards so close to his
chest.

'Wiles kept quiet for seven years about what he was doing,' said
Andrew Granville, an English mathematician at the University of
Georgia, 'and there was a reason for that. If he'd leaked some of the
things he'd done, the experts would have understood the implications,
so if he'd said, "Well, I've done this," some of the top experts would
have said, "Hold on a minute, if you could just develop this, this and
this," which was Andrew's plan, "then we have Fermat's Last Theorem."
And I know certain people who really feel that, had he not shut himself
up in his attic, they would at least have had a good go at it if he'd released
what he knew at the time he knew it. So some people are kind of
mad because he kept quiet for seven years about stuff he knew, and
other people were working on the same questions getting nowhere,
and this guy knew all this stuff.'

Of course, it might seem unreasonable to expect a mathematician who is within reach of the proof of the century – or one of them – to keep his colleagues fully up to date on his progress. But sharing findings is actually more reasonable than you might think in a subject like mathematics than it would be in, say, drug research or microelectronics. There is no immediate financial reward in mathematics – at least there wasn't for Fermat's Last Theorem – and science and maths have been traditionally open disciplines where every new result in a field is expected to be published as soon as it is verified so that other scientists or mathematicians can benefit from the discoveries in their own research. Wiles didn't want to work this way, which caused understandable distress to mathematicians who were on a similar track but further behind.

'They wasted a lot of time working on things that Andrew Wiles already knew,' said Granville. 'I think that they're not *so* angry with him. They understand why he did it, but they also feel that they wasted time when he already knew how to do this much better than they'd ever dreamed. It was upsetting for those other folks. We tend to be pretty collaborative.'

Like the Riemann Hypothesis, Fermat's Last Theorem had dominated the lives of some mathematicians to the point of obsession. Atle Selberg, the distinguished Norwegian number theorist, now in his eighties, has seen generations of mathematicians come to grief on specific problems, and indeed is believed by many of his colleagues still to be pursuing the Riemann Hypothesis after first reading about it in his teens and working seriously on it for the last sixty years.

'I grew up in a home with many mathematics books,' he began. 'My father had Riemann's collected works in his library. So I saw his paper and looked at it. My father had a degree in mathematics but he was not at an academic institution in that sense – he had a somewhat more varied career. He had taken up mathematics late, actually, but he never was near a library of any size so he decided he needed to have his own. We did not have Ramanujan's papers at that time, but one of my brothers had brought a copy home in the vacation. He had taken it out from university. And later my father bought me a copy for myself, which I still have. Seeing Ramanujan's collected works made a very deep impression. That was, I would say, what really started me off. My earlier papers were inspired by that. I was just about eighteen when I wrote my first paper.

It was "Über einige arithmetische Identitaten" – "On some arithmetical identities".

Selberg started working on the Riemann Hypothesis in 1940, and made a number of very important contributions to the theory of primes. But he never – in public – said that he was anywhere near a proof.

'You see, I have never had any idea that I thought could prove it. I have always ideas that could prove something else, that could prove some partial result, but so far I never saw any way. Actually, I think very few good mathematicians have any idea that they think could lead to a proof.'

But when I asked him whether he was still looking for 'the idea', he didn't actually deny it. 'Well, I've had to follow what is being done. I don't really expect now that I will have...' He hesitated and took a new tack. 'You see, as you get older you have the advantage of a great deal of experience, and you know a good bit, and you see there is also a good deal in your experience that you have not published. So you have a background of things that nobody else has in that sense. But there is no doubt that the big breakthroughs tend to come from the younger mind, that tends to have the more completely original ideas. As you get older you rely more on your experience, on your past...'

He thought for a moment and then, just in case I'd interpreted what he had just said to mean that he was past it, he added, 'My mind is still reasonably alert.' Selberg has probably spent more time than anyone else alive thinking about the Riemann Hypothesis, and there comes a time when you have to turn to something else for a while.

'You can't spend all your time at something like the Riemann Hypothesis. There were some people who spent their early lives on the Fermat problem. I think it was often the ruin of what could have been a more promising mathematical career. There was an American mathematician, Harry Vandiver, who was very talented when he was a young man. He got caught up in Fermat's Last Theorem, and spent his whole life on it. Of course, he did achieve some things, but it was always a little bit more of the same, going a bit further, pushing up the exponent for which you could prove it impossible, but without any way that seemed to point towards a solution of the whole problem. And of course the solution came from a side that had nothing to do with any of that.'

But Selberg did not share Granville's unease about Wiles having been so secretive – perhaps because Selberg's colleagues think he's being just as secretive with his own work on the Riemann Hypothesis.

'If you are working on something and you don't know whether it will pan out, so to say, then I think this is OK. You are not obliged to share everything before you would like to. Whenever you work on a mathematical problem it may be a fruitless pursuit. You may not succeed, or someone else may be working on it who succeeds before you. This is always a risk. And of course the bigger the problem, the bigger the risk, in a sense. I don't find any fault with Wiles for that. Some people would talk about it to some colleagues or friends while they are working on it; others, simply because of the notoriety of the problem, would keep more secret and not want it given out that they were trying to solve it.'

Henryk Iwaniec, another mathematician who is believed capable of proving the Riemann Hypothesis, also sees nothing wrong in keeping promising results to himself. 'I think I can risk a statement that anyone who has very promising idea would keep quiet,' he said, and laughed mischievously. 'I think when you hear people talking at conferences, that means either the results are premature or already digested very well and ruled out. Collaboration is a nice thing, but when it comes to the best work, this is too good to share.'

For Charles Ryavec, however, collaboration is sometimes the only way to do good mathematics. 'You need people around you. I was talking to one guy once and I said, "Suppose you're the last person on Earth: are you going to work on the Riemann Hypothesis?" And, you know, we kind of agreed "no". There's something funny about it: if you're the only one left, what are you supposed to do now? Turn off the lights? Shut the door? Do the Riemann Hypothesis? It's not going to be that important. I think I need other mathematicians around.'

'But what if you were among the last *five* people on Earth, all of them mathematicians?' I asked.

'Five or ten, I probably would,' he said.

The issue of priority of discovery – and in particular the credit for being the sole discoverer – is a very sensitive area for some mathematicians. Selberg was involved in an incident concerning the Prime Number Theorem which generated bad feeling among fellow mathematicians. Martin Huxley, at the University of Cardiff, describes it as 'a horrible mix-up'. The story concerned Paul Erdös, an eccentric Hungarian mathematician who memorably defined a mathematician as 'a machine for turning coffee into theorems'. Paul Hoffman, Erdös's biographer, describes the Selberg–Erdös spat as follows.

> The 1896 proof of the Prime Number Theorem [proved independently by Hadamard and de la Vallée-Poussin] depended on heavy machinery, and the brightest mathematical minds were convinced that that the theorem couldn't be proved with anything less. Erdös and Atle Selberg, a colleague who was not yet well known, stunned the mathematics world with an 'elementary' proof. According to Erdös's friends, the two agreed that they'd publish back-to-back papers in a leading journal delineating their respective contributions to the proof. Erdös then sent out postcards to mathematicians informing them that he and Selberg had conquered the Prime Number Theorem. Selberg apparently ran into a mathematician he didn't know who had received a postcard, and the mathematician immediately said, 'Have you heard? Erdös and What's-His-Name have an elementary proof of the Prime Number Theorem.' Reportedly, Selberg was so injured that he raced ahead and published without Erdös, and thus got the lion's share of credit for the proof. In 1950, Selberg alone was awarded the Fields Medal, the closest equivalent in mathematics to a Nobel Prize, in large part for his work on the Prime Number Theorem.[85]

'The most charitable explanation', said Martin Huxley, 'is that it was nobody's fault and that Selberg managed to dig a deeper hole by pretending he wasn't working on the Prime Number Theorem when he was. Perhaps Erdös had even been told how far Selberg had got, and Selberg had got this interesting-looking formula which he was fairly sure was a halfway step, and he was on his way to working out exactly what you do with it. Then, either he showed it to Erdös or Erdös had already had it leaked to him by a third person – the third-person version is the one that seeks to hurt either of them as little as possible. Maybe Selberg dug a hole for himself by saying, "I've got this formula, but I don't think it's good for anything," trying to put Erdös off the scent, not realizing Erdös would interpret this as having *carte blanche* to work on it. Erdös would probably have worked on it anyway – you can't take away knowledge.'

Selberg is a giant among number theorists alive today. He has worked on many of the classic problems, and contributed important proofs and techniques to the field. There is a Selberg Trace Formula and a Selberg Sieve among other eponymous products of his long career in mathematics. In fact, it was to consult Selberg that Hugh Montgomery went to Princeton in the first place, on the occasion of the now famous teatime meeting with Freeman Dyson.

Montgomery had been working on an analysis of the distribution of the zeros of the Riemann zeta function along the critical line. Odlyzko, in a paper about Montgomery's work, wrote:

> Relatively little work has been devoted to the precise distribution of the zeros. The main reason for the lack of research in this area was undoubtedly the feeling that there was little to be gained from studying problems harder than the Riemann Hypothesis if the Riemann Hypothesis itself could not be proved.[86]

Nevertheless, Montgomery did work on this topic, assuming for the sake of argument that the Riemann Hypothesis was true. 'Philosophically, I'm a firm believer in trying to form conjectures far beyond what one can prove,' he told me, 'because it's my feeling that the more you understand of the larger picture, of what lies beyond, the better idea you have about how you should try to prove even more modest things.'

By looking at a long list of zeros, Montgomery was able to calculate the average gap between pairs of zeros, and he found that small gaps occurred very infrequently. He came up with a function that described something called the pair correlation. This function said that if we analyse the gaps between the zeros, the differences between pairs of zeros obey a particular rule. (If you're interested, the rule is given by the formula $1 - [(\sin \pi u)/(\pi u)]^2$.)

Michael Berry, who was much later to find his own work transformed by Montgomery's ideas, described the significance of this function.

'He was able to see certain sums over prime numbers that he was able to analyse, miraculously, and he realized that this was interesting. He looked at the distribution of pairs of zeros and found an elementary function for the distribution which, most importantly, he quickly realized was not the answer that you would get if the zeros came down randomly. Usually, the philosophy in mathematics is your objects are either very structured or they come like random objects, and the heart of your problem is to prove that they do behave like random objects, and if they're random it means the structure's gone. So we like to think that they're either structured and we understand it, or they're random. And with the zeros which you numerically compute, you could make some statistical analysis to see if they're random numbers. Well, nobody had really looked at this kind of question, but Montgomery's analytic calculation proved that they can't be like random numbers, which was interesting, but he didn't know what they were.'

Odlyzko, who went on to test Montgomery's conjecture on some of the billions of zeros he calculated, described the significance of Montgomery's pair correlation conjecture.

'His results which assumed the Riemann Hypothesis suggested that the zeros would not be totally randomly spaced. They would not be like random numbers tossed on a line with no special properties, but that in fact there would be a particular distribution law for the spacings, in particular zeros would tend to repel each other, tend to stay fairly rigidly apart. It's probabilistic, so occasionally you get zeros close together but the result is that on average usually they don't come too close together. For example, if you look at the first billion zeros, if these things were just tossed uniformly at random onto the interval that they span then two of them would have a gap no bigger than about one in a billion. In fact, though, the smallest gap that you find is about one in a thousand – roughly the cube root of what it would be in the other case. So that gives you a quantitative measure of how the zeros repel each other.'

At this point, Odlyzko is clearly seduced by the analogy of the zeros as objects. 'Repelling each other' in a physical sense is not what he means, since 'forces' cannot act between abstract concepts like the zeros of a function of a complex variable. But it's a vivid way of describing the fact that the zeros have, on average, a greater distance between them than you would expect by chance.

Montgomery was intrigued by his discovery, but wasn't sure if it was original. So he decided to make a quick trip to the Institute for Advanced Study.

'I had spent the year before, 1970/71, as a visitor at the Institute and I had gotten to know Selberg. So I took the bus down to Princeton for the primary purpose of asking Selberg about these results, because at that time one never knew when one proved something whether this was going to turn out to be new or whether Selberg was going to say, "Yes it's been in my desk for years." And so I wasn't really guaranteed that this was a new development until Selberg said he hadn't seen it before.'

Montgomery laughed at this memory. 'I had also seen Freeman Dyson, who was in evidence around the community, but I don't know that he had the foggiest idea who I was, where I was from, what I worked in, even that I was a mathematician, so for all I know I was just another face in the crowd as far as he was concerned. And so I was back in Princeton for two days having afternoon tea in Fuld Hall, idling about.'

Among the other tea-drinkers that afternoon was Professor Sarva Daman Chowla, a distinguished Indian number theorist whose work built upon the foundations that Ramanujan had laid fifty years beforehand.

'Chowla had actually written a joint paper with Selberg,' Montgomery said, 'which is a testimony to his single-mindedness because Selberg is somebody who simply doesn't write joint papers with anyone. His only joint paper is with Chowla, who is this irresistible force meeting an immovable object, and the irresistible force won over Selberg. When Chowla would get into a certain mode he was really unstoppable.'

It was Chowla's unstoppability that led to the next step. Although Montgomery, a young researcher, would have had no specific reason for approaching the distinguished person of Freeman Dyson, Chowla thought it would be a good idea.

'He said, "Have you met Dyson?" and I said, "No," and he said, "I'll take you and introduce you to Dyson," and I said "No, no, that's OK, I don't need to meet Dyson." This went back and forth and it ended up with Chowla dragging me across the room. I didn't really want to bother [Dyson]. I didn't think of having anything useful to say to him, but when Chowla introduced me Dyson was very cordial and asked me what I'd been working on, and so I told him that I'd been looking at the zeros of the zeta function.'

It was when Montgomery mentioned the formula he had found for this distribution that Dyson's ears pricked up. At the mention of $1 - [(\sin \pi u)/(\pi u)]^2$, Dyson said something like, 'Well, that's the density of the pair correlation of eigenvalues of random matrices in the Gaussian Unitary Ensemble.'

'I'd never heard any of these terms before,' Montgomery went on. 'I don't know exactly what his words were because I have heard all of these terms many times since, but he said "pair correlation" and something resembling "random matrices"…'

What Dyson had spotted was a connection between two apparently unconnected fields of knowledge – quantum physics and number theory. It turned out that physicists looking for ways to characterize the behaviour of atomic particles had come up with a formula that was very similar to Montgomery's description of the zeros of the Riemann zeta function.

Peter Sarnak came much later to work at the Institute in Princeton, and by then the Montgomery–Dyson meeting had become part of mathematical folklore. Even if the details he heard – or remembers – differ from Montgomery's account, it is clear that, for him and others in the field, this was an occasion of almost mystical significance.

'Montgomery lectured on it here at the Institute in 1973, long before my time, and probably three guys turned up to his talk, maybe Bombieri, and Selberg, a few people who were kind of interested, but at tea he was introduced to Freeman Dyson, and Dyson said to him, "Young man," or something, "you lectured on the zeros of the zeta function – I couldn't come, perhaps you could tell me what you said." [Montgomery] started describing that he'd looked at the distribution of pairs in some scale, and Dyson said, "Did you get this answer?" That's roughly what I've heard – maybe it didn't exactly happen this way – meaning Dyson told him the answer and the guy said, "Yes, why?" and he said, "Because I've just been doing that calculation in the theory of random energy levels of Hamiltonians under certain symmetry, and that's what I get." And so to me, when I heard this many years later, this was a very compelling thing, because this doesn't happen by accident. And so it was a mystery why this was the case.'

This type of fortuitous connection of two seemingly unconnected pieces of mathematical knowledge happens from time to time in mathematics, and it's usually fruitful. There's a story about a New Zealander, Vaughn Jones, who was working on a problem to do with the field of 'infinite dimensional analysis'. Part of the answer involved something called the 'braid group', which can be represented as knotted or unknotted strands of a braid or plait. One day, at a conference in New York, Jones met a topologist, Joan Birman.

The French mathematician Alain Connes describes what happened next: 'Talking with Birman, he learned that the braid group is also used in knot theory, and that, as a consequence of a theorem due to Markov, topologists were looking for a function on this group that satisfied a certain property. "I've got it!" he cried. "I've got it right here in my pocket!"' [87]

For Montgomery, Dyson's observation was the first clue that he might have discovered something with wider significance than just the distribution of the zeros.

'I was very pleased because here is a conjecture about the differences within the zeros where there had been no conjecture at all before – no speculations, nothing. One had some reason to believe this would really be the way things would turn out, and yet this seemed to be unconnected with anything else. It seemed that there should be some interpretation that one should give to this that was lacking. This was great for me because it provided me with that interpretation.'

I asked Montgomery whether he'd done much up to that point with the mathematical entities mentioned by Dyson, known as random matrices. 'I'd never seen a random matrix,' he said. 'I've hardly seen one since.' Furthermore, for Freeman Dyson, this teatime exchange seems to have been just a momentary diversion from his own very different line of study.

'As far as I know,' Montgomery said, 'he didn't think about it after this five-minute conversation. I haven't spoken with him since, so I've had one conversation with him in my life. But it was quite a fruitful conversation. It was serendipity that it happened just at the right moment, because I had this result and what was needed was the connection, and he provided the connection. But it didn't alter the mathematics, it altered our understanding of what the mathematics was related to. I suppose that by now somebody [else] would have seen the connection…it's nearly thirty years ago. But it certainly was, from the standpoint of publication, instantaneous. I had the mathematics and as soon as I had it, it was just a matter of months before the connection was pointed out.'

From that conversation has come a whole new approach to the Riemann Hypothesis, and the possibility that in some quite significant way the quantum universe behaves as if it is driven by the location of the Riemann zeta function zeros.

10 A driven man

A mathematician's reputation rests on the number of bad proofs he has given.

A.S. Besicovitch [88]

In August 2000, Louis de Branges was waiting for a train at the small station of Gif-sur-Yvette, about fifty kilometres from Paris. While he was waiting, he sat and thought about proving the Riemann Hypothesis.

'That was a very inspiring time to think, and it seems to fit together. In fact, the key is a theory of the gamma function, something that goes back in my life to the time I was still in high school, a wonderful continuity. And so the theory of the gamma function is the answer. I have a procedure for going about it: I have to make calculations, and I'm confident that I simply have enough time to sit, to work it out, and the answer will be there.'

If it really was a key moment in de Branges's proof of the Riemann Hypothesis, the time he spent on the platform at Gif-sur-Yvette will mark the second major contribution of the French public transport system to the history of mathematics. The first was reported by the French mathematician Henri Poincaré at a famous lecture in 1913. He described how he had spent several restless nights trying to prove the non-existence of what are called 'Fuchsian' functions:

Just at this time I left Caen, where I was then living, to go on a geological excursion under the auspices of the School of Mines. The changes of travel made me forget my mathematical work. Having reached Coutances, we entered an omnibus to go some place or other. At the moment when I put my foot on the step the idea came to me, without anything in my former thoughts seeming to have paved the way for it, that the transformations I had used to define the Fuchsian functions were identical with those of non-Euclidean geometry. I did not verify the idea; I should not have had time, as, upon taking my seat in the omnibus, I went on with a conversation already commenced, but I felt a perfect certainty. On my return to Caen, for conscience' sake I verified the result at my leisure.[89]

Such moments are rare, but not *so* rare, as I discovered. Yoichi Motohashi had a 'Poincaré moment' in a bookstore.

'For seven years I had been thinking about the possibility of expressing the Riemann zeta function using some operator, but my solution wouldn't come. I live in the centre of Tokyo, and from time to time I go to a very nice second-hand bookshop. The atmosphere of that shop is close to some good bookshops in London. So I was in that bookshop quite close to my office, and I was looking at some beautiful book by the explorer Sven Hedin when suddenly the solution came to me! That is one of the most extraordinary experiences I've ever had. I started writing everything down, and sent the first draft to Matti Jutila,* and he recognized the value instantaneously.'

Enrico Bombieri, too, has had the experience of thinking intensively about a problem and then, unexpectedly and without warning, finding a solution. This time an Italian railway station features in the story.

'At the time I was working with Harold Davenport, a professor at Cambridge University in England, and we needed results on progressions. We obtained what we were aiming at, assuming what is called the extended Riemann Hypothesis, but then Davenport said that this could be done unconditionally, using large sieve techniques, provided that we could depend on some results in a particular paper. He was visiting Milan at the time – he was taking a week off to go in Italy as a tourist – so I said to him, "When you come in a week, I'll tell you what is really correct and what we can safely use." So I started looking at these things, and it became clear that in estimating certain sums there was an error. So I felt if maybe if I start with the sum over a rectangle, then I really get a sum of the same shape and I can feed it back. So I did a little calculation and it came out like a charm, and really in a few minutes I knew everything. I sat at my desk and worked for seventy-two hours, wrote the solution, met Davenport at the railway station, and gave him the paper.'

As I talked to de Branges in his apartment in August 2000, he told me how he intended to start writing what he called a background to his proof, which he thought would take some forty or fifty pages.

---

* One of Motohashi's close collaborators.

'I think that it could be two or three weeks for me to do the writing, and then it has to be typed; secretaries at the start of the semester are very busy with research proposals, so the typing could wait for a while. And then once the manuscript has been produced, I have to read through it and check a lot of things, and I think probably I'd be willing to take a semester or two to do that. I have no reason to hurry about it, and I would distribute it to at least one former student, and to [Nikolai] Nikolski, so that it would just sit, and be electronically available, until such time as somebody was willing to take it seriously. Now, it could be that it would catch fire.'

The proof itself was still in de Branges's head, but he was confident that he could write it out quite easily. 'I could maybe start to write it down on Wednesday,' he said. 'It could be as little as twelve pages; it might go to twenty, something like that. The previous theories were substantially more complicated. This is something that's rather simple. What I need to do now to prove the Riemann Hypothesis is to complete an axiomatic theory of the gamma function.'

But then de Branges did something which he was to do on several occasions in the course of our discussions about his 'proof' – he openly admitted his uncertainty. 'It could be wrong, you know,' he said. 'I just speak on the basis of my present situation, and there are times when things look good and times when things look bad. And I just do what's appropriate about it; I try not to care about it, although it has consequences for my life and for other people that are connected with me.'

To explain de Branges's approach to a proof of the Riemann Hypothesis is not easy. He works in a field largely of his own devising, called Hilbert spaces of entire functions. Not only is his method impossible to explain to a non-mathematician, but many mathematicians working in number theory or complex analysis would have to work quite hard on the theoretical background before they could understand the basic ideas.

In fact, the field emerged as a result of de Branges's early attempts to prove the Riemann Hypothesis, from the age of about twenty-five. But he had some difficulty then actually proving the relevance of this new field to the Riemann Hypothesis. He spent ten years on this task and finally, in 1984 in St Petersburg, he thought he had achieved it. He published this 'proof' in 1985, and then a pattern emerged which was fated to dog his attempts over the next fifteen years. The proof turned out to be wrong.

Peter Sarnak, who is now a key figure in the Riemann Hypothesis business, was one of many mathematicians who reacted against de Branges as a result of that first 'proof'. 'I was a young guy just out of graduate school, and I remember this manuscript arriving and some-body giving it to me. And nobody was looking at it. I actually personally felt that was crazy. I mean here's a guy who had solved this problem… and I spent many hours, and I found the big blunder. And it was *a big* blunder.'

Odlyzko also saw de Branges's early proof and was not convinced that even the approach was right. 'I understood what he was trying to do, but it did not seem hopeful in that it seemed that because he had a tool – Hilbert spaces of entire functions – it was like a big hammer and he was looking for a nail to hit it with. He was trying to manoeuvre the zeta function under this framework, and it didn't seem to be very natural.'

One of the surprising things about the working life of pure math-ematicians is the amount of time they can spend on a particular task. When you consider the intensity and narrowness of focus of mathematicians working on a deep problem, it is extraordinary that they can spend year after year with it as their preoccupation.

To spend a decade or more and then find you needn't have done so can be a major setback, as de Branges remembered. 'The problem is that a false direction may take ten years, and then you need ten years to come back, so you've committed twenty years to finding the truth. It's not been quite twenty years, but it's been somewhere in that direction.'

But undaunted (or rather daunted, but determined to continue), de Branges managed to find another way to tackle the proof, much simpler than his previous approach. He began to search for a type of mathemat-ical entity called an operator that would take another well-understood function and transform it so that all its zeros lay on the critical line. This distribution of zeros is called a spectrum, like the colours in a rainbow that run from red to violet, but instead of being smooth like a rainbow this spectrum is made up of many separate points. This spectral approach was something that Hilbert had predicted could work, and it is still the basis of other attempts, using the idea in different ways. But de Branges found that when he first 'operated' on the zeros, his new approach put all the zeros on a line a half a unit away from where they should be, on the edge of the critical strip. He either had to find a new operator, or modify the one he had. By 1989, he thought he'd cracked it.

He had the opportunity in the early part of that year to give a talk in Paris about his work, and he wanted to make the most of it. 'When I spoke I thought that I had a proof of the Riemann Hypothesis. And, you see, it's not possible to say, "I am doing interesting work on the Riemann Hypothesis – please listen to my work." People will simply not accept your relationship to it. So it's either that I have something which I think is a proof, or I don't.' He made his appearance, announced his proof, and then walked out, leaving the audience less than impressed.

Nikolai Nikolski explains why. 'He gave five lectures to explain the proof in Paris at the Henri Poincaré Institute in the summer of '89, and so I was not present because it was not a free era in the Soviet Union, but I heard about his lectures at the next conference in Amsterdam where I was for one week just the same summer. What I was told by someone who had heard these five lectures was that Louis spent the first four and a half lectures explaining the history of Riemann Hypothesis and number theory and then said, "Well, now we have half an hour so I will present quickly my approach." And he talked like this – block block block block block – until the proof was completed. This of course did not satisfy anybody. And to speak to him afterwards to ask him small technical questions is completely impossible. I tried myself several times, but he just says, "Oh, it's a very small, very easy calculation – you can do it yourself." It's just that he prefers to speak about principles and not enter into detail, and so it's disappointing, of course.'

There was worse to come. When de Branges returned to the United States he found a mistake in his proof. The problem was with the operator that was meant to shift everything to coincide with the critical line.

'If you have an operator,' he explained to me, 'it is some sort of black box – you're sending things in, you're getting things back; there's some mechanism for changing that black box, to move the spectrum over. And that mechanism works by shifting, and it's interesting that the shifting that's done is exactly one half-unit – exactly what's needed in the Riemann Hypothesis, because the distance, the critical strip, is one unit broad, and the critical line goes through the middle. And what you want is a spectrum on the critical line, and you have it one half-unit away. So you're going to have to push it over, and the conjecture that I have is a mechanism that does that. And my conjecture then was that you could just push it over, and that would prove the Riemann Hypothesis. Well, it turns out that life is not that simple. It's not that simple because the zeta function is a very complicated object. It has a lot of factors for all the prime numbers, and there are lots of prime numbers – an infinite number of them.

'And it took me ten years to convince myself that it couldn't be. It's only when I had spoken at the Institute that I had to go back and say, "How am I going to deal with all those primes there?" So in order to deal with all those primes, I needed to have a thorough theory of zeta functions, and I had to work it out. But it turns out, now that I've understood that theory and mastered it, that I don't need that information. And if you go about it right, you can put those primes in without any difficulty.'

At this point, de Branges was brought up against the fact that what he needed for the proof of the Riemann Hypothesis is something he'd already known in 1962. 'I simply had to see that that was sufficient,' he said, 'and that all the other information that I had discovered didn't help with the specific task of obtaining the proof of the hypothesis. The work I did on this theory, good as it was, was of no help to me whatsoever in my career, for promotion or for acceptance with my colleagues, because they weren't knowledgeable about it.'

The motif of hostility or neglect at the hands of other mathematicians is pervasive in de Branges's discussion of the evolution of his ideas. Whether or not this is paranoia, I certainly encountered a largely hostile reaction to de Branges from some other mathematicians, and when they weren't hostile they were irritated. Some had met him and been annoyed; others had formed their opinions just by reading what he wrote; still others seemed to base their hostility on hearsay.

Even the mild-mannered Michael Berry found de Branges a little hard to take. 'Isn't he strange? I spent about three hours talking to him in Paris. We met at a café – he said, "You'll recognize me by my beret." Of course, Frenchmen nowadays don't wear them, so you do recognize him. I'm afraid it was an unsatisfactory encounter, because all he wanted to do was to emphasize his priority and how people had stolen his ideas. Now, the point about him is that he's done one very good thing – he's proved the Bieberbach Conjecture. He's said a lot of things that turned out to be wrong, but one thing that turned out to be right, so you can't dismiss him. He's undoubtedly a very skilful analyst... His proof has something to do with operator theory, but it was impossible to talk about details with him. And he talked to me because I'm just a physicist, I'm low down, I'm no threat to him. So he was quite open, but he wasn't interested in anything that we were doing.'

Charles Ryavec – also a warm-hearted sort of person – had one encounter with de Branges which left him very sceptical. 'I haven't read Louis de Branges's stuff because I feel it would be a waste of time. I was at a talk he gave maybe fifteen years ago down in Anaheim, and he said he would solve [the Riemann Hypothesis] in the next year, just had to put a few ducks in a row and he'd get it, which is silly kind of talk. But people do rev themselves up that way – they put themselves on a deadline. I've read that a lot of writers couldn't write unless they put themselves on a deadline, had an editor ready to practically kill you or something and you had to get it done. So a year went by, and another year… He's a very good mathematician. I don't even call it a mistake. I think the only mistake of importance is in the proof, and I'm sure someone's going to read the darned thing and they're going to pass judgement on it. We are like a brotherhood… I think mathematicians would be perfectly willing to accept his proof if they thought there was something there. Personally I don't care who it is, they would read it.'

Unfortunately, Ryavec is being too optimistic. Reading a mathematician's proof of something as deep and difficult as the Riemann Hypothesis is not something you can do in an hour or two one evening with a coffee or a Scotch in your hand. And when the mathematician has already made two erroneous claims, you are likely to be even more reluctant to spend the time. There's one further factor too – de Branges is uncompromising in the way he writes mathematics, as he himself admits.

'When there is an innovation,' he said, 'it requires an effort on the part of the reader to master it. My conception of expression has two directions that many people would not accept. In the first place, I'm very Cartesian in my point of view. As a mathematician I'm interested in the logical sequence, and as a person that expresses myself in the use of English and in writing, the principal technique I have is being terse. In other words, I tend to choose exactly the minimal structure that's needed for saying what I want; I could be discursive if I had the time, but I don't. And the combination of the two directions means that a lot of people just don't like that way of writing. I'm expecting my reader to do a lot of work. For example, a comment that's been made about my book [on Hilbert spaces] by Russian readers – because the principal readers turned out to be in Russia – their comment is that the book has three hundred exercises in it, three hundred problems to be solved, and the principal content of the book lies in the exercises, so that I'm expecting the reader to work out these exercises without any hints to get the full meaning of the book. Well, it takes several years to do that. For doing mathematics it's essential for the student to work. Somebody can't tell you what it is – you're sent in the right direction, and you do the work. And it happens that there's a lot of work that has to be done, and I'm sorry about that but that's the nature of it, and so it's just a tough situation. And since the purpose of the book is not appreciated, essentially nobody makes that effort.'

Not surprisingly, a mathematical paper by a man who writes a book and expects his readers to take several years to work through it can be rather daunting. And in all my travels I came across only one man who seemed as if he might be prepared – albeit reluctantly – to assess de Branges's final version of the proof when it becomes available: Nikolai Nikolski, who helped de Branges on his proof of the Bieberbach Conjecture. Now at the University of Bordeaux, Nikolski is still in touch with de Branges, though he is rather wary of getting entangled in the time-consuming process of reading what de Branges writes. But he is one of the few people who understands how de Branges thinks, so de Branges needs someone like Nikolski if he is ever to persuade the world that he has proved the Riemann Hypothesis. But even Nikolski feels that de Branges's approach may not have the complexity required for such a tough proof.

'The Riemann Hypothesis is a problem of many faces,' he said. 'It depends on number theory and analysis and geometry. There is a link to non-Euclidean geometry, to many, many other things, so it's different. He has only *one* new idea – that this property should be explained in the language of some operator. And he just tried one operator after another: if it didn't work he replaced it with another, improving the argument, but I am not sure that this time it will be enough.'

When I asked Nikolski whether, nevertheless, he would look through de Branges's proof if he was sent it, he gave a sigh. 'You can't really "look through" a four-hundred page manuscript like, for example, the Bieberbach Conjecture proof, but at least I had that for two or three years. With

Louis, any assertion is either trivial and known, or incredibly complicated, and you've no idea how to verify it, no explanations. If you believe it will work, and if you have time you can spend it as we did with Bieberbach. But here, I don't know. Probably I'm older, or not so enthusiastic, but I have also many, many projects in process which have waited tens of years to be realized. I don't know...'

But when pressed, he had to admit that he *would* help with the task of assessing any de Branges proof of the Riemann Hypothesis. 'Yes, I will consider it. I know a little bit about the structure of the proof, so I probably will find the differences with respect to the previous manuscript and just consider them first from a general point of view. But really what he needs is some team of really enthusiastic high-level people. He found this with the Bieberbach proof in the middle of the eighties, in the only place in the world where there were a lot of very qualified people not touched by this pressure of the market to write something every year to publish in the respected journals. He found some curious people who just love to solve complicated problems and who were ready to spend a half a year to crack it even if it's not certain. He has asked me several times if it's possible to organize some people for his proof of the Riemann Hypothesis. I love him, so I said yes, we can do it if you have a very large grant, probably not so huge as in America, to pay some academic institution – for instance, the same place in [St] Petersburg – so I can try to invite some five to six people with good qualifications to be paid – because the salary now here is zero or almost zero – and to try to work hard to do this; but this proposal didn't get anywhere. In the eighties in the Soviet Union there were quite a lot of people who would agree to work just for curiosity. Now, unfortunately, it is simply not the case. Probably he had no money or was afraid that credit should be shared between the team – I don't know.'

De Branges's history of mistakes has made colleagues increasingly wary of trusting any new mathematics he presents. But mistakes, per se, are no bar to eventual success. Mathematical history is littered with examples of the most famous mathematicians who have nevertheless made – or been considered to have made – serious mistakes.

Georg Cantor, a German mathematician whose life and work straddled the nineteenth and twentieth centuries, was responsible for a number of ideas which shocked mathematicians at the time because they were so advanced. His work on different sizes of infinity, for example, was very difficult for the mathematical community to swallow at first. In 1885 Cantor wrote an article for the mathematical journal *Acta Mathematica*, but while it was being typeset he received a letter from the editor, Gösta Mittag-Leffler, pleading with him to withdraw it:

> I am convinced that the publication of your new work, before you have been able to explain new positive results, will greatly damage your reputation among mathematicians. I know very well that basically this is all the same to you. But if your theory is once discredited in this way, it will be a long time before it will again command the attention of the mathematical world. It may well be that you and your theory will never be given the justice you deserve in your lifetime. Then the theory will be rediscovered in a hundred years or so by someone else, and then it will subsequently be found out that you already had it all. Then, at least, you will be given justice. But in this way [by publishing the article], you will exercise no significant influence, which you naturally desire as does everyone who carries out scientific research.[90]

This rejection led Cantor to abandon mathematics and turn to philosophy, but his work *was* finally fully accepted, and he is now considered the founder of set theory – the very theory he had been warned not to publish.

In 1794, the French mathematician Adrien-Marie Legendre believed he had a proof of something called the parallel postulate. This was an axiom assumed by Euclid in his system of Euclidean geometry, and many mathematicians had tried to derive it from first principles (which we now know is impossible). Legendre was convinced it would be possible, as one text on the history of mathematics describes:

> In almost every edition of his *Éléments*, Legendre 'proved' the parallel postulate. However, each proof was attacked as insufficient by other mathematicians. With astonishing and stubborn persistence, refusing to consider the possibility that the parallel postulate might be false, he would provide a new proof in a later edition, hoping to satisfy his critics. In the third edition (1800) he replaced his original proof with one oriented towards the angle sum in a triangle. In the ninth edition (1812) he abandoned this proof, returning, as he later explained, 'to Euclid's simple way of proceeding, referring to notes for the rigorous demonstration.' He was particularly concerned with finding a proof suitable to insert in his *Éléments* for students to learn, and in the twelfth edition (1823) he

believed he had discovered the right one, which he kept in all the remaining editions published during his lifetime. (It was flawed, too, of course.) In a large memoir he finished in 1832, less than a year before he died, he wrote:

> The [postulate]…must be regarded as one of those fundamental truths which is impossible to dispute, and which are an enduring example of mathematical certitude, which one continually pursues and which one obtains only with great difficulty in the other branches of human knowledge.[91]

Stubbornly refusing to doubt the truth of the postulate, he even explained that an important reason for writing this final big memoir was that the proof he had given in the twelfth and all subsequent editions 'had not gained the agreement of certain professors who, without disputing its accuracy, had found it too difficult for their students to understand'.[92]

De Branges, too, has faced revolt from his students. Nikolski told me an interesting story about de Branges's early years at Purdue University: 'De Branges is the author of a textbook called *Square Summable Power Series*, and he survived a very sad story with this book because he had just invented this approach to analysis, and he started to teach it at undergraduate level at Purdue. But he was stopped by the administrators and sidelined from teaching at the undergraduate level after a massive protest by students. It was a big scandal, and so he was in a very strange position for two or more years. This tension continued until his proof of Bieberbach.'

The students found the course too intellectually demanding compared with the approaches of their other lecturers. But it is also likely that de Branges made no concessions to the fact that the field he was teaching would be strange and entirely new to his audience. Despite proving the Bieberbach Conjecture, de Branges, unlike many professors, has difficulty attracting students, and he puts that down to the attitudes of his colleagues.

'My ability for recruitment is seriously hampered by my situation,' he said, 'namely that [for] any student that I would have, I would have a responsibility for sorting out a career for that student, and my relationships with my colleagues at other universities are disastrous. They would not receive a student of mine. I've never refused a student, but what happens is that students with recognized competence expect that their teachers will run after them. And when the teachers don't run after them they are critical of their teachers. I require that the student chooses me as a professor, and is willing to accept the risk of the situation.'

De Branges is a driven man. He decided at an early stage in his career that he had a route to a proof of the Riemann Hypothesis, and he has never lost sight of that objective. This focus inevitably makes everyday life seem a little bit of a distraction. I get the sense that journeys and meals and social chit-chat and taxes and television are all dispensable in the face of the demands on his brain for a next stage in The Proof. Yet, from time to time there are signs that he is in contact with the outside world. He surprised me one day by his familiarity with the novels of Agatha Christie, and then it turned out that there was a puritan purpose behind such pleasure-seeking. He had wanted to learn German, and decided to achieve this by reading Agatha Christie novels in German. On another occasion he reminded me that Professor Moriarty – Sherlock Holmes's nemesis – was an expert on the binomial formula.

At the age of sixty-eight, by which time many mathematicians would have settled contentedly into retirement, de Branges laid out his hopes and fears for the future. 'The status of my work is at this moment undecided. Mathematicians usually are classified by the age of forty, whether or not they get a substantial prize. To get a Fields Medal you can't essentially be over the age of forty, at least for the work that's done, and here in my case the decision about the work that I did before I was thirty has not yet been made. So, you see, it's quite an unusual situation.'

With a display of what was either quiet realism or hubris, de Branges described what would happen if he failed to achieve his aims over the next few years. 'I would give the impression of being somebody that had made a lot of fireworks that would impress society, but I wouldn't be someone that would have directed mathematics in a new channel, or maintained the channels of the twentieth century. It wouldn't be the realization of mathematics, or of functional analysis in the twentieth century.'

At the time of my first visit to de Branges he was on the point of putting the final touches to his proof of the Riemann Hypothesis. I planned to see him next later in the year, at Purdue University. Would the proof be complete by then? Would the hapless Nikolski have his head buried in a long and difficult manuscript? Or would the proof be as near – and as far – as it had ever been?

(11) The physics of mathematics

There's been an explosion of activity in this field – the progress in
the last half dozen years because of this marriage of these two fields
has been absolutely incredible.

Steve Gonek

It's a sign of the depth of the Riemann Hypothesis that the deeper mathematicians go into the topic, the wider the network of connections they
uncover. The search for a proof moved away a long time ago from what
is called elementary number theory into other fields of mathematics,
some devised specifically to attack the problem. But no one could have
predicted that a proof might be found outside mathematics – in physics
– and that in some still unfathomable way the Riemann zeros bear an
uncanny resemblance to the behaviour of hydrogen atoms in a very
strong magnetic field.

Physics and mathematics have been known for millennia to be intertwined. Pythagoras observed that the lengths of vibrating strings that
produce harmonious sounds are in integer ratios ($\frac{1}{2}, \frac{1}{4}, \frac{1}{8}$, and so on), so
that the harmonious interval of the octave is produced by two strings,
one of which is half the length of the other. But no one has ever been
able to relate the prime numbers to any physical system – until now.

The new insights are emerging from a decades-long study of the
behaviour of atomic nuclei, work carried out by nuclear physicists to
analyse how atoms behave under different conditions. In particular,
physicists have been interested in charting the huge variety of states,
called energy levels, that can be associated with individual atoms or
atomic particles. Until the twentieth century, our understanding of the
movement of matter – from peas to planets – under the influence of
forces was governed by what's called classical mechanics, based on the
laws described by Isaac Newton. But as physicists explored the recesses
of molecules and atoms, it became clear that the tiniest atomic particles

behaved in a *non*-classical way, and the new tools of quantum mechanics were devised to describe them. Scientists moved from describing the behaviour of individual particles or other objects to looking at the characteristics of whole collections of particles and describing them statistically, almost in the way that census data can describe a society.

More recently, over the last thirty years or so, the even newer field of chaos theory has been developed to explain an unpredictable type of behaviour that occurs in physical systems, both large and small, under certain conditions. This wasn't initially applied in the field of quantum mechanics, but then it was realized that in certain circumstances even the smallest particles sometimes behave unpredictably, and needed a new theory to describe them.

The applied mathematics developed to deal with these new insights into atomic physics has almost taken on a life of its own. A mathematical technique called random matrices is used to handle thousands or millions of pieces of data generated when quantum mechanics is applied to a system of particles, and it is beginning to look as though the results of doing this can produce data which are suspiciously similar to the Riemann zeros. The idea that links all these topics together along a path that leads to the Riemann zeros began at that tea party at which Hugh Montgomery met Freeman Dyson.

At the time of their meeting, Montgomery had never heard of random matrices. Nowadays, in some quarters of Riemannology, as Michael Berry likes to call the field, they talk of little else. A mathematical technique used to handle data in atomic physics has turned out to have unexpected implications for number theory. This is not the usual way things happen when these two disciplines interact. There are many stories of abstract mathematical ideas which have proved to be surprisingly useful in physics or chemistry, years – or centuries – after their discovery. But this marriage of quantum mechanics and number theory is a much rarer event.

This marriage has brought about a collaboration between Professor Michael Berry and his former student and now colleague, Professor Jon Keating, of the University of Bristol. Keating is the maths half and Berry the physics, but each has become very familiar with the other's subject. Keating explained to me how the ability to deal with the statistics of randomness is an essential tool in modern physics.

'The notion of randomness is fairly ubiquitous in the natural sciences', he said. 'If I want to compute the pressure on the wall at the moment due to air molecules hitting it, I could do that by finding the position of all the air molecules, solving Newton's equations, finding out the times they hit the wall, and computing the pressure that way. Instead, it's viewed as sensible to assume that the air molecules are moving in all random directions – which they're not. There are 10^{23} air molecules in the room, and they're moving in perfectly well-described trajectories if you believe Newtonian mechanics. But the numbers could be statistically indistinguishable from a random sequence.'

Keating has helped Berry devise mathematical tools to describe what are called in physics 'quantum chaotic systems', and it turns out that applying those tools can lead to a mathematical function that looks remarkably like the Riemann zeta function.

Michael Berry works in Bristol University's H.H. Wills Physics Laboratory, named after a cigarette manufacturer in the days when cigarettes were believed to do no more harm than drop ash on your pinstriped suit. When I visited him first, in March 2000, the sunlight was streaming unseasonably into his small cluttered office, and with no concern for one's preconception of how a knight of the realm should dress, he was wearing bedroom slippers. Of medium height, with a head of clipped hair and a beard of similar length, Berry is an amiable, fast-talking man, who sometimes leaves words half-finished in his rush to get to the next idea. And the ideas come thick and fast.

I was intrigued by the idea that new mathematics could emerge from physics. I couldn't see how the kind of experimental work that I thought physicists did could offer much more than problems for mathematicians to solve. There are all sorts of phenomena that physicists discover that require some form of existing mathematics to describe. Sometimes, as in the case of one of Ramanujan's findings, a piece of mathematics that seems to have no practical application can turn out years later to be just the job for explaining a phenomenon that could not have been discovered in Ramanujan's time.

But Berry's work, and that of his mathematical colleague Jon Keating, was different. Somehow, it seemed, his understanding of quantum chaotic systems – Berry calls this study 'quantum chaology' – had led him to a remarkable idea: that the Riemann zeta function behaves as if there is an underlying dynamic system controlling the position of all those zeros. This would not mean that an actual physical system was driving the Riemann zeta function – how could that be? – but that it might be possible to find a mathematical description of a perfectly feasible physical system, that, if it existed, would throw up the Riemann zeros as energy levels or as some other physical characteristic.

Berry is not a mathematical physicist, or a physicist-mathematician; he described himself as working on 'the physics of the mathematics of physics'. In trying to understand this gnomic expression, there was a lot to take in. But an hour or so later, I thought I had grasped the bare bones of Berry's ideas about the Riemann Hypothesis, and I could see what the phrase meant.

As Berry explained, 'There's a lot of mathematics that comes out of physics: differential equations, matrices and algebra, for example. I study the solutions of these mathematical equations that relate to physics by having pictures of the mathematics. Take wave equations [a key technique in quantum mechanics]. They come from sound waves, from light waves, from quantum waves. But once you've *got* a wave equation, the mathematics is just the same, and when I come to understand the patterns in these equations I think of it in terms of the physics of waves but not in terms of any *specific* physics, like whether it's sound or light or quantum or whatever, because a lot of the concepts are at a higher level. But they're still physical concepts which come out of the mathematics that comes from a particular type of physics.'

As a physicist, Berry studies the mathematics of the behaviour of atomic particles. Their movements, energies and other characteristics are described by numbers. Since it's rarely possible to observe and measure these characteristics for a single particle, the data that physicists gather are statistical, derived from the characteristics of many millions or billions of particles.

The discipline of quantum mechanics, which underpins our modern understanding of the microscopic world, was put together in the 1920s, building on discoveries and insights from preceding decades. One of them was a description of how electrons orbiting the nuclei of atoms behave when they gain energy. When you add energy to an atom, you don't get a smooth increase in the atom's energy. Instead, what happens to the electrons is a little like pushing objects upstairs, one step at a time. If you start with a series of weights resting on different steps on a staircase and subject them to slowly increasing upward forces, the weights could end up on higher steps. If a particular force isn't strong enough to lift a weight up to the next step, the weight will stay put. A very strong force might lift a weight two or three steps. But in any case, the end positions of the weights would always be on steps, never hovering at some intermediate position between two steps.

Or, in an analogy Keating gave me, 'In classical mechanics, energy can take any value you like, so if you're in your car and you want to give it a little bit more energy, you put your foot on the gas a bit more. In quantum mechanics that's not the case, so that the energy that the electron in an atom can have takes particular discrete values.' With atoms, there are many thousands of steps, or orbitals as they are called, that can be occupied by the atomic particles when raised to higher energies.

Over the second half of the twentieth century, scientists developed a detailed understanding of how some of these collections of atomic particles behaved, using the insights of quantum mechanics. But there were some types of behaviour and groupings of atoms that didn't seem to obey the rules of quantum mechanics. Several atoms bound together in a particular way, or individual atoms subject to a combination of forces, behaved in a way which physicists found difficult to predict. But then along came chaos.

Chaos theory was developed to describe macroscopic systems – those made up of much larger elements than subatomic particles – which behaved in a way that should have been predictable but wasn't. An example of a chaotic system is a small iron sphere on a string, like a pendulum, suspended between two magnets (call them green and red). If the sphere is held at some starting point and then released and allowed to swing freely, sooner or later it will stop in a position where it is pointing towards one or other of the magnets. But it's very difficult to choose a starting point that will *guarantee* that the sphere will finish up nearer the red magnet, say. However closely you try to make the starting point the same on each try, it only needs the tiniest discrepancy, and thus the tiniest of changes in the initial magnetic forces acting on the sphere, to make the sphere end up near the green magnet and not the red one.

'The message of chaos,' said Berry, 'is that a system doesn't have to be complicated for its motion to be complicated. That's what chaos is all about.' So the field of *quantum* chaos has emerged, where ideas of chaos are used to try to get a better understanding of the motions and energy states of certain systems of atomic particles. Out of this synthesis of quantum mechanics and chaos theory has come a mass of data which demands new tools for it to be analysed. Those tools are random matrices.

Like a charged and highly excited particle himself, Berry sets off on a tour of the subject, bouncing from one topic to another, occasionally dropping to a lower energy level or leaping up to a higher one. Although he is a physicist, he is rarely to be seen doing experiments, in this field anyway. But at some point the data he works with came from real hands-on physics.

There have been hundreds of experiments designed to probe atomic structure by firing neutrons – particles from the nuclei of atoms – at different targets, usually blocks or sheets of some substance. Even apparently solid matter contains a lot of space between the atomic nuclei, but sometimes a neutron fired through matter will hit a nucleus or pass near it and be deflected. The way the neutrons pass through, bounce off or are slowed down by the target tells you something about what's called the cross-section of the atomic nuclei in the target. This cross-section is a kind of average of the nuclear forces that repel or divert the neutron. By the 1950s, nuclear scientists had accumulated vast banks of data from neutron capture cross-sections which were needed for nuclear reactors and other technological purposes. They were just sets of numbers with no explanation attached, but could perhaps be analysed statistically for more information about the atomic particles.

'Because you deal with very high excited energy levels,' Berry explained, 'it's often not very interesting to know exactly where each level is. What is interesting is the statistics of how these numbers are arranged – whether they repel each other, whether they're regular, whether they're random, and so on. Now we got the idea from the nuclear physicists who had been doing it in the fifties and sixties. Their idea was that if you got the 10,000th excited state of some nucleus, you're never going to get a quantum-mechanical theory which will predict that exactly, because you don't know the forces well enough. But you can ask about the statistics. There, the problem is complicated because you have many particles.'

The task facing physicists was to find a way of extracting useful data out of millions and millions of individual pieces of information. It clearly cried out for a technique which would handle lots of data at the same time, and that's where matrices came in.

A matrix in mathematics is just an array or grid of numbers which can be manipulated as if it were a single number. So if you have two matrices *A* and *B*, they can be multiplied together or added to each other to make another matrix. When the individual elements in matrices represent the physical states of atoms or nuclei, the act of performing mathematical procedures on the matrices can correspond to physical processes that change a subatomic system from one state to another.

Berry and his colleagues believe that a collection of matrices associated with the chaotic behaviour of certain systems of atomic particles may have characteristics that are similar to the collection of zeros of the Riemann zeta function. These matrices are called random matrices, and they can have millions of elements. Associated with each matrix is a set of numbers called eigenvalues. Sometimes these eigenvalues are complex numbers, but when they are used in physics they represent real quantities such as energy and momentum, and so they are real numbers. The fact that the eigenvalues are real numbers also means that you can arrange them in a single sequence in numerical order, to produce what's called a spectrum. (For more on matrices and eigenvalues, see *Toolkit 6.*)

Berry and Keating have analysed their data – the spectra of eigenvalues – and compared the analysis with the data Andrew Odlyzko produces from the Riemann zeta zeros. They find great similarities. It's almost as if the Riemann zeros themselves are like physical entities.

'The Riemann zeros fit firmly into the class of chaotic systems,' said Berry, 'and this is a powerful hint that underlying the Riemann zeta function there's some unknown dynamical system which you could think about as classical dynamics, some chaotic trajectories – no one knows what it is, even how many dimensions it has, but if you then pretend it's quantum mechanics, do physics with it, then the energy levels of that system will be the Riemann zeros.'

For a time, it seemed that random-matrix theory was enough to describe the statistics of quantum energy levels of classically chaotic systems. But when Odlyzko and Berry first started swapping ideas, Odlyzko found that when he looked at some of the larger zeros as if they were eigenvalues of some dynamic system, thing started to go wrong.

'Odlyzko found these numbers and he was very worried,' Berry said, 'because they disagreed with our numbers, in a kind of way that you'd expect if you'd made some little mistake somewhere.' But then he and his colleagues found that a new understanding of the physics meant that they needed to change *their* numbers. 'In 1985 I made the beginnings of the explanation for the connection between random-matrix theory and quantum chaology. This demonstrated that quantum-chaotic systems are described by random matrices only over short ranges of energy levels – long-range correlations are different, revealing that a quantum-chaotic system cannot be completely modelled by a random matrix. Just at that time I was working out this theory of how you'd get deviations if you consider ever-larger groups of energy levels, and I was able to apply it to the Riemann zeros. So I said to Odlyzko, "Can I have your data?" And I adjusted it and got this nice smooth curve which was beautiful. It is these characteristic differences that I predicted would also appear in the distribution of Riemann zeros – precisely the differences revealed by Odlyzko's computations. The very strong suggestion is therefore that the Riemann zeros are eigenvalues not of a random matrix, but of a matrix corresponding to a quantum system whose classical dynamics is chaotic.'

Berry's excitement is infectious, but he's not there yet.

'Many people have tried to solve this problem,' Peter Sarnak said, 'and everything becomes circular after a while. If I can prove that this is equivalent to this, there's no input. The philosophy that most people have nowadays is that you really have to understand these objects in a way where you'll get some new information from somewhere.' But Sarnak believes that random-matrix theory could well produce new information, *if* – and it's a big if – it's possible to describe a physical system whose energy levels are the Riemann zeta zeros. Now this needn't be a system that actually exists, it only has to be one that *could* exist.

The American physicist and polymath Murray Gell-Mann has applied to physics a phrase from T. H. White's *The Sword in the Stone* that 'everything that is not forbidden is compulsory', by which he means that if the physics allow the existence of some system or process, it will exist somewhere in the universe. While Berry does not believe he's anywhere near finding a physical system that behaves in a way that generates the Riemann zeros, he wouldn't rule it out.

'What I believe,' he said, 'is that there will be some mathematical dynamical system describing a particle in a space. This would be a mathematically generated thing, but because it has a structure of a classical dynamical system it could be implemented by some suitable experiment – with magnetic fields or in optics or whatever – but wouldn't come naturally from the world. This [system] would have the property that if you looked at it in a refined way [by using] quantum mechanics, it would have energy levels, and those energy levels would be the heights of the Riemann zeros. It isn't that you would look with a spectroscope some-where, and see the Riemann zeros coming as spectra, but that some clever person will be able to construct a system, a physical system, that may seem very artificial at first but will get to seem more and more natural.'

It would be quite a coup if the most important unsolved problem in mathematics were to be solved by a physicist rather than a mathemati-cian, and not many mathematicians think this is likely. But, as Berry said to me, 'We can dream. Keating and I talk about it… We exhausted our-selves two or three years ago, and wrote up what we know so far about it in a long review article whose last section includes a particular conjec-ture, but at the moment I sort of doubt it.'

However, there *is* someone who may well build on Berry's and Keating's work in a new mathematical way that might lead to a proof. This is Alain Connes, a French mathematician who came much later to the Riemann Hypothesis than the other major players, but now has the bit between his teeth.

Berry described him to me as 'one of the world's greatest mathemati-cians, very imaginative, and he has this idea of what the dynamical system is. He's lovely. There are some mathematicians who are very re-ceptive in the sense that they realize that we physicists don't prove things, but we often know things or guess things in ways that they don't, and then they can develop them in directions we wouldn't ever dream of. It's really two-way traffic. I don't see any hierarchy in this at all, but some mathematicians are very scornful. I used to say that I'm optimistic. The optimist believes that we live in the best of all possible worlds. The pessimist *knows* that we do. But now I'm not quite so optimistic. I think Connes may have found it, and it will be a question of understanding it, but I'm not quite sure yet. We're far from that at the moment.'

I planned to see the 'lovely' Alain Connes, though after a whirlwind trip through Berry's mind I wasn't sure how much more abstraction I could take. But to be fair, even Berry finds that he cannot think in these higher realms for twenty-four hours a day.

'This kind of stuff is in our bones,' he said, 'but it's utterly mysterious if you haven't studied it for a number of years. It's the kind of thing that can drive you mad, and we find that we're obsessed with it every few years for a few weeks, and we make some little progress every time we do it, but then we exhaust ourselves and we do other things, and maybe that's why we won't succeed and why Wiles did... I have this question: "What is the elementary particle of sudden understanding?" It's the "clariton". The problem is there are also anti-claritons that come tomorrow and annihilate the one you had today. We have lots of them in this subject, claritons followed by anti-claritons.'

A laudable aim

Some decades ago I made – somewhat in jest – the suggestion that one should get accepted a non-proliferation treaty of zeta functions. There was becoming such an overwhelming variety of these objects.

Atle Selberg, 2001

The Riemann zeta function extends its tentacles into so many branches of mathematics that it's impossible to forecast from where – or from whom – a proof of the Riemann Hypothesis will come. I assumed that Michael Berry moved in Riemann Hypothesis circles, and that he might have a sense of who had a chance of proving it.

'The thing is, there aren't circles,' he told me. 'There are isolated people. The Riemann Hypothesis is a big thing, and some people will be motivated by personal ambition, and they would work quietly away and not tell anybody, as Andrew Wiles did with the Fermat theorem.'

Brian Conrey believes that 'isolated people' working on the Riemann Hypothesis is a bad idea. Tall and relaxed, and looking young for his age, early forties, he lives in a large low house in a residential community to the south of Silicon Valley. As we talked he pointed to a building on a hill overlooking his house.

'That used to be a restaurant,' he said, 'then there was a wedding reception out on the raised deck and it collapsed, leading to a lot of lawsuits. We're hoping that it will become AIM's new center for mathematics.'

The AIM – the American Institute of Mathematics – was founded by electronics millionaire John Fry to encourage the type of cooperation that mathematicians working on the Riemann Hypothesis sometimes shy away from. Conrey, the director of AIM, is himself tackling the Riemann Hypothesis but, like everyone else who is suspected of trying to find a proof, he says he doesn't have a strong idea. But he does have half an idea which, in the AIM collaborative spirit, he was quite happy to talk me about.

'The Riemann Hypothesis is the most basic connection between addition and multiplication that there is,' Conrey told me, 'so I think of it in the simplest terms as something really basic that we don't understand about the link between addition and multiplication.' That connection comes in the beautiful formula discovered by Euler, where a series of terms involving all the integers are *added* together and shown to equal a series of terms involving the primes that are *multiplied* together. 'I think most number theorists believe you really need to use both these ingredients to find a proof for the Riemann Hypothesis,' he said.

Conrey's idea for a proof uses an unusual function called the Möbius function. It's unusual because it brings home how arbitrary the idea of a function is, and yet how useful it can be. On the whole, the functions we've looked at so far can be plotted on a graph that varies continuously as the variable increases or decreases. A function such as $f(x) = x^2$ increases smoothly as x increases; $f(x) = 1/x^2$ *decreases* smoothly as x increases, apart from right at the centre, between negative and positive values of x, where $x = 0$ and the function shoots off to infinity. But the Möbius function is far from continuous. In fact, it only ever assumes three values, depending on certain characteristics of x. It is either +1, −1 or 0. So the 'graph' of the function is just a series of points scattered in an apparently random fashion between those three positions as x increases.

The Möbius function of a number n is denoted by $\mu(n)$ and applies only to whole numbers. The symbol μ (pronounced 'mew') is the Greek equivalent of the letter 'm'. Here's how it works. If n is a prime number, $\mu(n)$ is defined as −1. If n is a composite number, created by multiplying several primes together, there are three possibilities: at least one of the primes is repeated, in which case $\mu(n) = 0$; or there is an even number of different primes, in which case $\mu(n) = +1$; or there is an odd number of different primes, in which case $\mu(n) = -1$.

So what is $\mu(15)$, for example? Well, the Fundamental Theorem of Arithmetic tells us that 15 can be expressed as a product of prime factors in only one way: as 3×5. No prime is repeated, and there is an even number of different primes, so $\mu(15) = +1$; $\mu(29) = -1$, because it's a prime; $\mu(30) = \mu(3 \times 2 \times 5) = -1$, because 30 is made up of an odd number of primes multiplied together; and $\mu(96) = \mu(3 \times 2 \times 2 \times 2 \times 2 \times 2) = 0$, because at least one of the primes is repeated.

This may all sound rather puzzling – what is a function like, if it adopts these rather odd values in this way? But to think this way is to miss the point. The function is like this because a mathematician (August Ferdinand Möbius, famous for his one-sided strip) *defined* it to be so in 1832. I could define a function and call it the Sabbagh function, $S(n)$ and make it do anything I wanted. I could, for example, say that $S(n)$ equals 5 for all even values of n and 117.382 for all odd values of n. Or maybe $S(n)$ could equal $\sqrt{n}$ whenever n is a perfect square, and $\frac{1}{2}$ if it isn't. These would be perfectly valid functions – the only thing is, they would be of absolutely no use to anyone, whereas, presumably, the Möbius function was defined in this odd way for some purpose.

The values of the Möbius function for the numbers 1 to 10 are:

$$1, \ -1, \ -1, \ 0, \ -1, \ 1, \ 1, \ 0, \ 0, \ 1$$

and the series continues in an equally arid way.

Associated with the Möbius function is another function called the Mertens function, $M(n)$ for short. To find $M(n)$ you add up all the values of the Möbius function from 1 to n. So if we look at the values of $\mu(n)$ from 1 to 10 we can see that the values of $M(n)$ from $n = 1$ to 10 are

$$1, \ 0, \ -1, \ -1, \ -2, \ -1, \ -2, \ -2, \ -2, \ -1$$

Believe me when I tell you that, in spite of appearances, this is *much* more interesting than the Möbius function, for one simple reason: if the absolute value* of $M(n)$ can be proved to be always less than the square root of n, then the Riemann Hypothesis is true. This is called Mertens's conjecture.

To see how it works, take $n = 10$. In the list above, the absolute value of $M(10)$ is 1, which is consistent with the Riemann Hypothesis being true, since the square root of 10 is about 3.1622. This conjecture was shown to be true for very high values of n, and for many years it looked as if it would turn out to be true for all values, though no one could prove it. Then along came Andrew Odlyzko and his colleague, Herman

---

* The absolute value of a number is the number itself if it is positive and the number without a minus sign if it is negative. The absolute value of n is written $|n|$, and so $|-n| = |n| = n$. $|n|$ is called the modules of n – 'mod n' for short.

te Riele, and they showed in 1984 that there is a number, somewhat larger than 10^{30}, that invalidates Mertens's conjecture – call it N. In other words, $M(N)$ is greater than the square root of N. So the conjecture is not true. But, in the frustrating way that can happen in maths, this is not a proof that the Riemann Hypothesis is false. While a proof of Mertens conjecture would effectively have proved the Riemann Hypothesis, a *dis*proof of Merten's conjecture does not *dis*prove the Riemann Hypothesis.

Conrey's interest in the Möbius function has come about because you can use the series of values of $\mu(n)$ up to infinity to write an expression that is equivalent to the Riemann zeta function. As he described his approach, even though the maths was highly specialized, I got a sense of how this particular mathematician worked away at a problem and of the excitement that could come from each small advance in the process.

'There are a whole bunch of different steps in the theory, and a lot of stuff to know just to get into it. Then, once you've got it you're tinkering around with it in a lot of different places. There's definitely an intuitive feel to it, there's a bit of an art to it. You have formulas and lots of checking and playing with different parameters and seeing how they worked, but you also have your gut feeling, your instinct for how the thing's going to go. Putting all these things together, you know that you're going along the right lines without having to check everything, and then you see that you get there, and then you go back and see that everything's absolutely correct. And so all of that requires a lot of skill, intuition, experience, this kind of stuff. It's kind of scouting things out in a way, looking for a good trail.

'On the one hand, it seems pretty clear that this approach is not going to get you to the Riemann Hypothesis. On the other hand, I do have some ideas about a mechanism where this could conceivably lead to the Riemann Hypothesis, but I haven't been able to get those worked through. I think most people probably wouldn't give much credit to this particular approach, so I would say I would be the most optimistic person you'd talk to about this approach leading somewhere. I'm hoping for a really beautiful relationship involving the Möbius function to somehow magically appear. So it's magic.' He laughed at the outrageousness of the metaphor.

Conrey's insight into his own thought processes was matched by that of other mathematicians I spoke to. They each had a slightly different way of creating the best conditions for that moment of inspiration, the 'magic' that Conrey sought. Sometimes just knowing or believing that a solution to a problem exists can spur on a mathematician to find it. Roger Heath-Brown once heard that someone else had solved a particular problem.

'I wondered how this author can have proved this, and in musing on it, within the space of half a day I'd come up with a key idea that would enable me to solve this problem. It was one of those days when you get a balloon above your head with a light bulb in it. "Aha, right, goodness me!" And then there's a hard slog, probably uninspiring. I find that a very time-consuming process. I've never in that process come across something I can't put right. It's like being a washing-machine repairman – you know your machine and you can always put it right. Maybe you have to replace more and more of the parts, but you can always put it right.'

Charles Ryavec pointed out that one of the things that made maths difficult for the professional mathematician trying to break new ground is that you never know how near you are to your goal. 'I remember I wanted to be an athlete,' he said. 'If you wanted to be a high jumper and you were going to go to the Olympics, you'd say, "OK, I've got to jump eight feet." So you go out and you work very, very hard, and maybe you jump five feet, and you put in another year and it's five foot three – you actually see the gap closing. In mathematics it isn't that way. You can't really see where you're suffering or how far it is to the solution. That is the real difference between an intellectual pursuit and a physical one – in mathematics, you just don't know. It may be that some brilliant mathematician has worked very hard on the Riemann Hypothesis, but he cannot decide if he passed within an inch of it or within a million miles, and then he's going to feel frustrated and he just won't even want to talk about it. When they went up Everest, they could say, "OK, we have another thousand feet this way and, OK, maybe we have to carry oxygen tanks." You can figure it all out in a funny way that's pretty standard, it's just stamina. But *maths is not just stamina*. There's a lot of that in it, but there's this other intangible. I was watching the biography of [Harry S.] Truman last night on television, and the guy was a failure most of his life – just one constant failure – but he kept going, for some reason.'

Littlewood wrote about a key moment in his proof of what is called the Abel–Tauber Theorem:

One day I was playing round with the derivatives theorem, and a ghost of an idea entered my mind of making r, the number of differentiations, large. At that moment the spring cleaning that was in progress reached the room I was working in, and there was nothing for it but to go walking for two hours, in pouring rain. The problem seethed violently in my mind: the material was disordered and cluttered up with irrelevant complications cleared away in the final version, and the 'idea' was vague and elusive. Finally, I stopped in the rain, gazing blankly for minutes on end over a little bridge into a stream, and presently a flooding certainty came into my mind that the thing was done. The forty minutes before I got back and could verify it were none the less tense.[93]

But whatever the delights of the solitary mathematical enterprise, it's sharing ideas with others that can take a mathematical problem in new directions. A key event in the Riemann Hypothesis story in recent years was a conference organized by AIM and held in Seattle in 1996, on the hundredth anniversary of the proof of the Prime Number Theorem. Attended by five hundred mathematicians, it created an upsurge of interest in the random-matrices approach, and led Conrey and his colleagues to plan smaller meetings – workshops – so that they could go into some of the ideas that were emerging in greater depth. One of these workshops was held in Palo Alto in May, 2001. Its subject was 'L-functions and random matrix theory', and it provided a good opportunity for me to observe a group of mathematicians, most of whom knew each other well, as they rolled up their sleeves and did real mathematics at the frontiers of their field. One of the participants, Andrew Granville, felt that such workshops are valuable, though he was sceptical about whether they could lead to breakthroughs.

'I think John Fry is a little optimistic,' he said. 'Team efforts have not in history been the way that great breakthroughs have happened. Two people working together certainly, for example Montgomery and Vaughan. Many of the great results in the subject are due to Montgomery *and* Vaughan together. Certainly, Hardy and Littlewood at Cambridge. But what works very nicely is an exchange of ideas. And most people have got the idea that this is fairly open, and they try to have a fairly positive attitude about things that they may not really believe in. But I think more ideas are good. It's good to hear ideas and take them in and see where they lead. But whether this is really a big step on the road to a proof of the Riemann Hypothesis, Lord alone knows.'

When Fry set up the AIM, he thought that if he spent a few hundred thousand dollars flying the best people in the field to one place, putting them together for a week and telling them to prove the Riemann Hypothesis, then they would. It's an attitude similar to Kennedy's pledge to put men on the Moon by the end of the 1960s – which worked – or Nixon's to cure cancer by a similar onslaught of brain power – which didn't. Fry's initiative was likewise unsuccessful.

In spite of his disappointment that the Seattle meeting didn't produce a breakthrough, Fry felt strongly that collaboration could still work, but would take a little longer. The 'L-functions' workshop was attended by a motley group of mathematicians, from Atle Selberg, described to me by one overawed participant as the most brilliant mathematician in this field alive today, to fresh-faced individuals barely out of short trousers. In fact, *in* short trousers, since the dress code seemed to permit every kind of clothing from formal suits to near-underwear. There was a burly Australian in denims who looked as if he could have stepped off a building site – but was actually a professor at Berkeley; an Indian who would have been a dead ringer for Ramanujan; a Chinese mathematician who taught at Brigham Young; a scattering of Englishmen and -women; an Afro-American; a red-headed Israeli from Tel Aviv; a lugubrious cryptanalyst from New Jersey wearing a baseball cap; and the benignly avuncular Brian Conrey, who observed from the back and ensured that there were constant supplies of coffee, sodas, bagels, muffins, and bottles of good red wine and Mexican beer.

The workshop began on a Monday morning with a display from Nick Katz, a collaborator of Peter Sarnak's, dressed casually in short-sleeved shirt, chinos, socks and sandals, with light grey hair and a dark Groucho Marx moustache. There was a lot of hand-waving as he described the Katz–Sarnak Conjecture, a statement that, if true, may turn out to be a medium-sized contribution to the proof of the Riemann Hypothesis.

'There are four guys here,' said Katz, pointing to a crudely drawn circle on a whiteboard, and he started to show how 'the guys' pair up. 'If there's one here, there's one there; and if there's one here there's another one there.' The 'guys' in this case were eigenvalues of random matrices. Since they were complex numbers they couldn't be arranged along a line. But they could be placed around the circumference of a circle, which is what Katz was doing.

The Katz–Sarnak conjecture is a statement about 'L-functions', which are mathematical objects that arouse Brian Conrey's enthusiasm: 'It's a whole beautiful subject and the Riemann zeta function is just the first one of these, but it's just the tip of the iceberg. They are just the most amazing objects, these L-functions – the fact that they exist, and have these incredible properties and are tied up with all these arithmetical things – and it's just a beautiful subject. Discovering these things is like discovering a gemstone or something. You're amazed that this thing exists, has these properties and can do this.'

Sarnak told me how L-functions are a type of zeta function. 'We have this zoo of zeta functions, including L-functions, and Riemann is just the first guy born in the zoo. And everything we know about the Riemann zeta function we know for the zoo of zeta functions. There's a whole bunch of them – they are the most important objects in number theory, and whatever we know about zeta we know about these other ones. And just as we can't prove the Riemann Hypothesis for the Riemann zeta function, we can't prove the Riemann Hypothesis for any of these other ones.'

Jon Keating tried to encapsulate the essence of L-functions for me. 'All their formulae are similar,' he said. 'They're all an infinite sum, not of $1/n^s$ but of *something* over n^s. And they all have a Riemann Hypothesis. Also, extremely importantly, they can be written as a product over prime numbers, and that's absolutely crucial to this business because there are some objects which can be written as sum of $1/n^s$ but they don't have a Riemann Hypothesis. And that's because they can't be written as a product over primes. People have studied the L-functions individually. They all look like the Riemann zeta function, they all have a critical line where the zeros are believed to lie, there are heights of those zeros and you can analyse them statistically.'

We can begin to see why Atle Selberg might have wanted a non-proliferation treaty for zeta functions. It's difficult enough getting one's head around the complex four-dimensional landscape that is the *Riemann* zeta function. In the same four-dimensional space we now find that there is an infinite number of other surfaces, all of them dipping down to sea level, where their zeros lie, but dipping down at different places from one another. Nevertheless, the belief is that, like the Riemann zeta function, all the L-functions' zeros also lie on the critical line.

Instead of looking at a large number of zeros in one function, the Riemann zeta function, Katz and Sarnak compared a few zeros from whole families of functions. They plotted a sequence of numbers which was, in a sense, at right angles to the conventional spectrum of zeros, by taking the first zero of each L-function (see Figure 14).

'What they found,' said Keating, 'was that there are some strong indications of remarkably simple hidden symmetries in all the families of L-functions that you look at. So, for example, every L-function, of which the Riemann zeta function is an example, has a first zero. It's just a number, so to each L-function there's a number, and you now look at a whole variety of L-functions, a family, and analyse those numbers for that family and do the statistics on those. They found that, depending on which family you take, you get different answers, and those answers have a very simple form and are related to extremely simple groups that appear as very simple geometrical objects. That's observation number

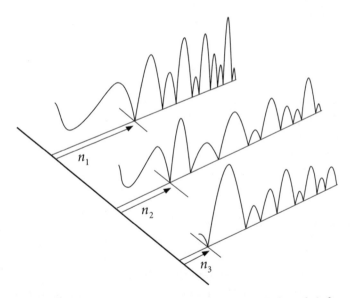

FIGURE 14 Three L-functions, represented very symbolically, have their first zeros at different points (numbers n_1, n_2 and n_3). If we look at a whole group of L-functions in this way, we can list the positions of all the first zeros as a series of ns and thus analyse the properties of that series.

one. Observation number two is that the family that the Riemann zeta function naturally falls into has, as its symmetry, one of the simple ones, and it's actually the symmetry that appears in classical mechanics.'

But what exactly are these L-functions, and how are they connected with the Riemann zeta function? The connection is through prime numbers. As we've seen, the Riemann zeta function can be used to calculate how many primes there are below any number you care to specify. As a variant on that, the L-functions deal with how many primes *of a certain type* there are up to a given number. One L-function relates to primes of the form $4n + 1$, for example. Numbers of this form are not all prime, but some of them are: $4 + 1 = 5$ (prime), $8 + 1 = 9$ (not prime), $12 + 1 = 13$ (prime), $16 + 1 = 17$ (prime), $20 + 1 = 21$ (not prime), and so on. Thus there are three primes of the type $4n + 1$ among the integers up to 21, whereas there are eight primes altogether. You can think of many different structures for numbers that might be primes, and there's an L-function for each of them.

As Katz got into his stride, his scribbling on the blackboard got faster and his hand-waving increased. And hand-waving is exactly the right word here. It has two meanings in the world of mathematics. First, of course, it refers to Katz's expressive actions as he covered two large whiteboards with symbols, sketches and graphs. But more fundamentally, hand-waving describes a process that is sometimes an important stage in developing an argument in front of colleagues. In thinking through a series of mathematical steps, mathematicians will not always proceed methodically. If they feel fairly sure that a particular step is justified, but requires a certain amount of hack work to establish, they will skip ahead to the next step, then the one after, where more interesting thinking may lie. There may well be a bigger obstacle ahead that destroys their argument anyway, so why waste time now on work that could turn out to be pointless? Normally, when their work comes to be written up and presented at formal conferences, all the intermediate steps will have been filled in with the necessary detailed reasoning. But here, in front of whiteboards being wiped clean and filled again for the umpteenth time, Katz is among friends – like-minded mathematicians who are happy to accept hand-waving as a key part of the argument. They don't want to sit through more minutes of basic maths than is necessary when they can be speeding ahead with Katz into the new territory of low-lying eigenvalues, the subject of the Katz–Sarnak Conjecture. Like him, they know that hand-waving, particularly in the early stages of an argument, will almost certainly be replaced by accurate proofs at a later stage.

During the lunch break in an oriental restaurant, nine mathematicians, including a couple of Englishmen, indulged in shop talk – the appalling state of salaries and conditions in British universities for mathematicians; the likelihood, or improbability, that the Riemann Hypothesis would be solved in anyone's lifetime; bets that had been won or lost that the Riemann Hypothesis would be proved.

'Hey, I've just remembered,' shouted some man down the table, 'Brian Conrey owes me a bottle of wine – he bet me it would be solved by the end of the last millennium!'

When the bill for lunch came, one mathematician was deputed to work out how much each share was to be. 'I need to divide by nine,' said a professor of mathematics from a mid-western university. 'Divide by three and then by three again,' shouted a voice from down the table, and the group dissolved into laughter.

Andrew Granville drew my attention to a small spat that had taken place during a talk by Henryk Iwaniec. 'You may not have heard because it was rather dead-panly done by Iwaniec, but one thing that's rather annoying is when you think through a topic and come up with some rather interesting thoughts that may be relevant to a big question, and afterwards somebody says, "I thought about that fifteen years ago and pretty much came to the same conclusion." And so you say, "Where did you publish it?" And he says, "Well, I didn't." And you say, "Where did you talk about it?" And he says, "Well, I never have." And you say, "Well, OK, that was your choice, but you can't lay credit to something that nobody else knows about." In Henryk's talk he just said, "I've heard other people have done something like this, but since it hasn't appeared in print or anywhere I know about, I feel I have a right to develop the results in my own way." And it was a poignant remark, because it's clear that some people have kept ideas of a similar nature in the drawer for a long time. That's their lookout – they were hoping they were going to get the big one out of it or some version of the Riemann Hypothesis – but Henryk's pretty much convinced me that this is a dead end, and he's writing a paper to explain why it's a dead end, and it's a very valuable paper.'

Granville also told a strange story about a similar thing that had happened to him the year before. 'For the first time ever, I thought I had an idea for proving a weak form of the Riemann Hypothesis, and I put it in a grant proposal to the National Science Foundation. One of the anonymous peer reviewers wrote back and said, "Oh, this idea's been tried years ago by Roger Heath-Brown and shown not to work." So I saw Roger at a conference when I presented my results on the subject – I had developed it and realized why it wasn't going to work, but actually I found some interesting things nonetheless – so I said to Roger, "What did you do on this? Somebody told me that you'd got somewhere on this," and he said, "Oh, yes, I had this idea some years ago but I didn't really get the results." He's quite honest, Roger. He said, "I didn't really get the quality of results you got, but I had the fundamental idea and I realized it wasn't going to work." And I said to him, "How public was this?" And he said, "Well, I was once going to a conference in Germany and I bumped into two analytic number theorists on a train, and I remember I told them about it in the train carriage, but I don't think I told anybody else." So it was annoying that a referee was almost accusing me of plagiarizing Roger's idea on the basis of a conversation in a train carriage fifteen years ago!'

The 'closed-door' syndrome is clearly something that worries mathematicians all the time. Working at the frontiers of mathematics is difficult stuff, and it is worth anything at all only if what you are doing is unique. Every painter who paints Mont Saint-Victoire produces a unique and valid work of art. It may not be very good, but it is in no way diminished if someone comes up and says, 'Oh, Cezanne did one of those already.' But maths is different.

'I think what happens with things in the drawer,' said Granville, 'is that you persuade yourself of something but you don't work out all the details, and then when you work out the details things pan out differently. I had this situation once: I gave a lecture in Princeton, and afterwards Selberg came up to me and said [here Granville put on a Norwegian accent], "Ven I first proved your theorem in 1954, I had thought about it like this," and so he explained to me what he'd done and it was a far better perspective than I had, so it was clear the guy was completely straight up about it. He'd kept it in the drawer waiting for the right use, and hadn't published it. But the extraordinary thing about his mind was that thirty-five years later he could remember every detail. I can't remember what I did *three weeks* ago. Who knows what people are doing behind closed doors?'

13 'No simple matter'

I am sure that Louis de Branges's many 'wrong' proofs of RH and other conjectures are as chuck-full of brilliant ideas as is his proof of Bieberbach.

Doron Zeilberger [94]

In a letter to me dated 28 September 2000, Louis de Branges wrote, 'The proof of the Riemann Hypothesis is taking longer than expected. I have a revised strategy for the end game and aim for a complete manuscript in October.'

I had decided to go to Indiana to see him on his home ground, hoping to get a sense of a university professor at work in a busy academic environment. The time that suited him, however, was Thanksgiving, when the only people around on campus are orphans, foreigners and curmudgeons who don't celebrate the major American festivals. He paid little attention to Thanksgiving, and could therefore spare the time to see me.

De Branges banned me from hiring a car at Indianapolis Airport – 'You won't know how to find your way to Lafayette,' he said, though I have been finding my way by car from strange airports to new destinations in the US for thirty years. I allowed him to meet me, and after taking an hour to find our way out of Indianapolis he drove me to Lafayette, sixty miles away. We travelled on minor roads to avoid the freeway, which we sometimes crossed, sometimes drove parallel to, but always appeared to be carrying freely flowing traffic. It became clear as we talked on the journey that even if I had visited him during term time I wouldn't have seen much more academic activity on his part. 'The holiday started on Tuesday,' de Branges said. 'I gave a lecture on Monday.' 'How many people were at your lecture?' I asked him. 'Well,' he said, 'I only have one student at the moment, an Indian.' In my university days, a lecture consisted of a professor talking for an hour to a hall full of students. I had heard that in American classes there were often many hundreds of students – too many, sometimes, for effective information flow. I was to learn more about de Branges's much smaller classes when I sat in on one the following day.

Purdue campus at Thanksgiving was like a ghost town. We arrived at about eight o'clock in the evening, and drove past darkened multi-storey parking lots and university department buildings, almost indistinguishable in the gloom. At the heart of the campus there was a faint prospect of nourishment in the neon signs of bagel shops and coffee houses. Some even had signs that said 'Open' but were firmly shut. Shadowy figures slipped around corners, huddled close to the wall. Most were foreigners – Chinese, Indian, Arab – students who, presumably, were too far from home to leave for the holiday, a holiday which in any case meant very little to them.

The following day, Thanksgiving itself, de Branges and I walked in the bright winter sunshine from the university hotel to his office in the Mathematical Sciences building. Vast stretches of paving lay before us, handbills and posters pasted to the flagstones here and there, but there wasn't a human being in sight. A nearby carillon rang out with premature Christmas carols.

As we passed the parking lot that de Branges used, he told me proudly of the day when, as a Distinguished Professor, he was given a reserved space. 'It meant I could go home to lunch,' he said. 'Up till then I couldn't leave the campus at lunchtime because there wouldn't be a parking space when I came back.'

De Branges has been at Purdue since 1962 and feels there is nowhere else to go, even though he works in solitary isolation, suspicious of colleagues who might otherwise be collaborators. Even his students (he uses the word to mean people he once taught who are now professors in their own right) no longer support his views, offering themselves as commentators on his work to people who should really be coming to him.

I had lunch at de Branges's house, and met his wife Tatiana and his stepson Kostia. They seem alone in an American suburb, their lifestyle constrained by the demands of de Branges's never-ending self-imposed task. Their television set receives only one channel, and they rarely watch anything but the news, at 6.30 p.m., a regular ritual.

'We cannot afford more television,' de Branges said. It was time, not money, he couldn't afford – valuable time that would take him away from his mathematics. Tatiana is a painter, with no mathematical interest at all. She used to work in a department of astronomy but stopped when the maths became too difficult. What must it be like, I wonder, to live with Louis, for whom there is nothing in the world but mathematics and persecution?

Today is a crucial time for de Branges's assault on the Riemann Hypothesis. He has reached a point where a key conjecture, one that has to be proved before he finishes writing his proof, is resisting his attempts to prove it. With a hypothesis such as Riemann's, which is difficult to state simply, it's even harder to convey the flavour of the approaches to a proof taken by different mathematicians. But what de Branges is doing that nobody else thinks is viable is concentrating on a function called the gamma function, which has played a part in his life for almost fifty years.

'I heard about this function when I was still in school, in my next-to-last year,' de Branges said. 'I knew there was such a function, and I tried to find it and I did find it. I knew that it existed, and I had some help from a book and then I actually derived it. I thought I'd made an achievement. You have to be very careful as a student – you can think you've made an achievement but really it was known already. It influenced my later work.' The gamma function was important in de Branges's Bieberbach Conjecture work, and what he learnt then has led him to see it as a crucial part of proving the Riemann Hypothesis.

This function is an example of something that happens often in mathematics and can be a fruitful source of new ideas. It's a process we've seen in Toolkit 1, when mathematicians asked the question, 'What would a number with a fractional exponent look like?' There was no obvious way to interpret $9^{1/2}$, because it seemed that powers only made sense with whole numbers. If 9^7 was 9 multiplied by itself seven times, what was 9 multiplied by itself a half times? But reasoning and extrapolation led to an interpretation of fractional powers, and then decimal powers, and then even imaginary powers. These were not so much discoveries as inventions – mathematicians chose to call $n^{1/2}$ the square root of n (also written as $\sqrt{n}$) because it worked.

The gamma function sprang from a similar simple origin. There is an expression called the factorial of a whole number. The factorial of n is written $n!$, and it stands for the result of multiplying together all the positive whole numbers that are n or less. So, factorial 6 equals $6 \times 5 \times 4 \times 3 \times 2 \times 1 = 720$ and is represented by 6!. The gamma function, written $\Gamma(n)$, is a mathematical expression that sometimes behaves in the same way as the factorial expression, in that for whole numbers $\Gamma(n) = (n-1)!$ So,

$$\Gamma(7) = 6! = 6 \times 5 \times 4 \times 3 \times 2 \times 1 = 720$$

The function itself is constructed in such a way that you can substitute fractional values or even complex values for n and get meaningful answers that turn out to have many uses in analytic number theory. For these other values of n, the gamma function is written as

$$\Gamma(n) = \int_0^\infty x^{n-1} e^{-x} \, dx$$

What looks like a long s ($\int$) after the equals sign is what's called an integral sign, and is just another mathematical symbol that acts like an instruction to do something to the letters and numbers that follow it. In this case, it's a type of sum (that's why the $\int$ looks a bit like an s).

In devising new functions that extend the use of a simpler function beyond the whole numbers, there are a couple of important conditions that must be fulfilled. First, it goes without saying that if you feed the *whole* numbers into the new function, you must get the same result as with the simpler function. So if you substitute 7 for n in the integral form of the gamma function, the answer should still be 6!, in other words 720. Second, the same relationships should apply between different values of the function with non-integral n as they do with values where n is a whole number. With whole-number factorials, $n! = n \times (n-1)!$. You can see that with the example of factorial 6, for which $6! = 6 \times 5!$ (in other words, $6 \times 5 \times 4 \times 3 \times 2 \times 1 = 6 \times (5 \times 4 \times 3 \times 2 \times 1)$). For the gamma function, it should therefore also be the case that $\Gamma(n) = n \times \Gamma(n-1)$. It's this function, simple but profound – like the Riemann zeta function itself – that de Branges is convinced will supply the proof. He sees himself as continuing a particular tradition in number theory, travelling a path towards the Riemann Hypothesis that no one else nowadays thinks is a viable approach.

'It's an implementation of a classical conception,' he said. 'In 1880, examples were given of functions that had the desired zeros. These are constructed from the gamma function factor alone, so that's the main motivation for the Riemann Hypothesis. It's the direction that…Hardy [favoured]. Furthermore, as far as modern mathematics is concerned, I am competing with people using scattering theory, or using spectral theories. Within these fields my own spectral theories are very powerful, and my competence in these fields is of a very high level, and it's simply a question of writing it out and getting the publications there.'

But in fact, by November 2000, it wasn't quite 'simply a question of writing it out'. De Branges still had to modify a proof of a conjecture that he needed in order to establish his full proof of the Riemann Hypothesis.

'I have a very promising modification,' de Branges told me, 'and in order to confirm that modification I have to make a calculation, [but] I haven't made that calculation yet. It's something that could go quite quickly, but until I've made that calculation I have to say that the nature of the Riemann Hypothesis is that it's all or nothing. In other words, no partial result is going to be convincing to anyone and it would be very difficult to publish. There's also now the problem that there's a reward, and so I wouldn't want to make information available that another person could use.'

Louis and Tatiana invited me to supper, and since I had been banned from hiring a car Louis had to come and pick me up. I knew how important it was to de Branges to watch the news every evening, so, since they could only watch CBS at home, I suggested they came and watched the news in my hotel room, where I had thirty-seven channels. Tatiana and Louis turned up at 6.20 and we settled down in my room. I started channel-hopping.

'Which would you like?' I said. 'CNN, ABC, NBC …' – offering them a selection of other news bulletins.

'No,' said Louis, 'we'd better watch CBS. We always watch CBS. Any other bulletin would disrupt our continuity.' Tatiana had no say in the matter. So we all solemnly sat watching a news bulletin we could have watched in their own house.

Over the meal, we talked about what he will do about media interest when he wins the prize. 'People will want to interview me,' he said, 'but I won't be interviewed. They will say what they want anyway. They will go to other mathematicians, people who in the past have accused me of unreliability, and print what they say. And if I win the prize, people will think I am rich and it will be impossible to get funding for future work. Everyone will get the wrong idea.'

'But, Louis,' I said, 'if you are even 95 per cent right about the fact that journalists will be more interested in what your rivals will say against you, what have you got to lose by talking to the media? They will talk to your rivals anyway. If you give interviews too, there is at least a chance that your views will also be printed.'

'It will be too distracting to give interviews,' de Branges replied. 'It will stop me continuing with my work.'

'But if you've proved the Riemann Hypothesis won't you want to stop for a while anyway?'

'No, of course not. I have so much more to do…'

On Friday morning I walked from the university club, where I was staying, to the Maths Department. De Branges was planning to give a class during which he would work on the conjecture he needed to prove. His one student was an Indian from Bengal named Yashowanto Narayan Ghosh, 'Yasho' for short, whose main subject of study was statistical mathematics but who had a love of the kind of number theory that de Branges practised. It seemed that de Branges had only ever had one or two students at a time. This surprised me at first until I read about similar situations in the history of mathematics. Isaac Newton frequently came to the lecture-room and found it empty. He would wait for a quarter of an hour, and if no one came he'd return to his rooms. Jack Littlewood remembers being the only student at a lecture by his tutor, E.W. Barnes, on his current work on double gamma and zeta functions. When the Russian mathematician Yuri Matijasevich was working on a proof (eventually successful) of Hilbert's tenth problem, he had difficulty retaining the interest of other mathematicians: 'The first meeting where I gave a survey of known results was attended by five logicians and five number theorists, but then the numbers of participants decreased exponentially and soon I was left alone.'[95]

On Friday, 24 November 2000, Louis de Branges and Yasho were the sole occupants of a classroom in which there were seven blackboards spread around three walls. De Branges did most of the talking, while writing in a neat, clear hand on the boards. Although the content was impenetrable, it looked like algebra, something I can still manage to grasp. Expressions were divided into other expressions and multiplied by yet others. Some of the symbols were unfamiliar, but it still looked vaguely like one of the nightmare problems I used to get in school where each stage of working out got you deeper and deeper into complexity, when what you hoped for was that it would simplify as you went along.

From time to time de Branges stopped, stood back and looked at something he had just written, apparently feeling uneasy. Yasho chipped in to point out an error. The dialogue during one of those moments conveys the arcane flavour of the proceedings:

'... and we have $z-\alpha$ and $\bar{w}-\bar{\beta}$ common,'* said de Branges as he wrote, 'we'll put that over a common denominator so $z-\bar{w}$, $\alpha-\bar{w}$ will be down there. In the numerator we have to have $(\alpha-\bar{w})-(z-\alpha)$...there is going to be a cancellation of the alphas...'

'Is $z-\bar{w}$...that is the different...?' said Yasho, hesitating while he peered at the blackboard.

'Yes, there is...' said de Branges, understanding perfectly Yasho's incomplete sentence. 'There's an error. z... This should be $\bar{w}-\bar{\beta}$. No, wait a second...'

'It is all right...'

'Something is not working right...$z-\alpha$, $\bar{w}-\bar{\beta}$...?'

'Is it a different factor in the first term?' Yasho wondered. 'Is $z-\bar{w}$, so that should appear in there...?'

'This is $\alpha-\bar{w}$, $z-\alpha$ ... right, I've copied them right.' De Branges still sensed there was something wrong. 'But, oh, here, down below...'

'Yes,' said Yasho.

'... down below I've made an error, $z-\alpha$, $z-\alpha$...Oh, it's not working out the way I expected.'

'No, it is, the error was in the numerator.'

'Oh, this is, this is...'

'The denominator was all right...'

'OK, $z-\bar{w}$, so...'

'The second term in the numerator – that should be a $z-\bar{w}$...'

They seemed to be back on track. 'OK, now it's looking good,' said de Branges, 'the $\bar{w}$ s cancel, and the $z-\alpha$ will come out with a minus sign...'

What surprised me about this kind of rarefied maths in action was its mixture of inspiration, perspiration and chalk. Even a mathematician of the calibre of de Branges has to make sure that his brackets are all in the right place, that the signs change when he changes sides and that he doesn't write alpha when he means beta. There was a further complica-

---

* $\bar{w}-\bar{\beta}$ is pronounced 'w bar minus beta bar'.

tion the day that I was there in that, though he had seven blackboards to write on, he eventually ran out of space and had to erase the earlier stages of the calculation, which meant that he couldn't check back when he needed to. Sometimes he had to transfer a complicated expression from one blackboard to another, being careful not to rub something out that he would need to copy. It was all very primitive. These are not things that can be easily word-processed or even computerized. There was a lot of room for a slip of the chalk, and Yasho was careful to watch out for these.

After an hour or so, de Branges had reached a point of stasis. He'd arrived at an expression which needed to be put into other terms and then substituted in an equation, but he wasn't sure how best to change it so that it would work.

The process of thinking out loud provided good insights for Yasho: 'I feel that this style of teaching is very effective,' he told me. 'This is the way in which the European master painters used to teach – the master used to work, and the students used to learn by watching.'

I began to understand what de Branges actually does when he is working on the Riemann Hypothesis. In some ways he is indistinguishable from an undergraduate working on an elementary problem in differential equations: paper on his lap, sharpened pencil, a string of symbols, some crossings out, a check back to see that there were no obvious errors, a puzzled gasp when the answer you expect to be, say, $\pi/4$ emerges as an awkward and cumbersome decimal that has no relation to the answer in the back of the book, and with a sinking feeling you know that something has gone wrong. When such things happen to de Branges – and such a thing did happen in Thanksgiving week, 2000 – there is no book with answers in the back. There may even be no answer at all. The real world of mathematics is far removed from that of maths professors who set their students neat problems. De Branges's fear was that, having got as far as he had in the last twenty-five years, this might be the end of the road.

When the class was over, de Branges said to Yasho, 'A calculation has been made. The question is now to see whether the results of the calculation can be adapted or meet the expectations. The expectations are that a certain kind of argument can be put through. That argument requires certain objects to exist with certain structures, and I'm looking to see whether they are actually there. And you see, from the point of view of the Riemann Hypothesis, the zeta function is an extremely complicated expression simply because there are so many prime numbers. All the rest that's been done is a lot of preparation – it's a lot of thinking – it's a lot of saying, "What do we want to do?" and "What is the route for going about it?" What we are talking about today is the only step where there's something substantial to do… This little algebra here, that decides the whole Riemann Hypothesis – you happen to be here at a very opportune time.'

On our way to lunch, de Branges said to his wife: 'Tatiana, I'm going to take Karl on a drive around the town. He said he'd like to see the old buildings of Lafayette.'

'We could do that now,' said Tatiana, 'or we could do it after lunch.'

'No, we should do it now,' said de Branges, 'I need some time to change my mood before we have lunch.'

If he really had reached a roadblock in his solution, his mood cannot have been good. But whether a drive round downtown Lafayette would be enough to change it was questionable. But it seemed to work. He wasn't exactly cheerful over lunch, but a kind of cheery resignation hung about him, the feeling that he probably would be able to prove the conjecture that was at the heart of his solution, but if not, hey, he had enough problems with his university colleagues, the costs of his household, the lack of funding for his research – enough for one more obstacle in a project spanning most of his life not to matter too much.

As he looked back over the path that had brought him to today, he resorted to a mountaineering metaphor that mathematicians often use. 'This moment is really a moment of elation, you see. It's like a manic depressive: there are times when you have a high and there are times when you have a low, and when you're on the high side you even it out, and when you're on the low side again you take it easy. In 1985 there was a conjecture that implies the Riemann Hypothesis, and there was a short cut that was not right. But on the other hand, the idea that you could go from one point to another had some merit to it – it's just that you had to go a little farther. Then in the summer of this year, another route was found – it seemed promising but again it's a short cut. We realized in the fall – and Yasho was part of that – that the expectations we had for a route simply were unreasonable: they weren't compatible with the existing situation, and it was necessary to have a more complicated route.

In other words, if we're trying to climb Mount Everest, the last bit of the way is very narrow – we don't have choices about how to get there, and this seems to be the only route, the only route that will get there, if the whole conception is correct.'

Although de Branges has seen enough moments of depression followed by elation, he is always ready to concede that this time things might not work out. 'We have the possibility that the conception from the start, going back forty years, is inadequate. What the majority of other mathematicians would say is that this is all nonsense from the start, so it would simply confirm their expectations if we fail.'

It was the sort of situation that once made the Hungarian mathematician George Pólya say, 'Those who want to climb the Matterhorn ought first to visit the graveyard in Zermatt.'

In May 2001 I decided to phone de Branges. I'd had the odd letter from him, one enclosing a copy of an autobiography of his mother, but we hadn't spoken for a while and I wondered how he was getting on. He was in France for his regular summer trip, and as we talked he told me that his proof was going well. Apparently, during my Thanksgiving visit to Purdue, he had indeed cracked the problem that had been worrying him. We discussed the possibility of another visit by me to Gif-sur-Yvette. This time, I suggested, maybe I should come for a shorter time, a day or so. 'No, stay longer,' he said, 'that way you can reach into the back of my mind.'

It was a mind I still wanted to reach into the back of. It was now almost a year since I had first come across de Branges, a year in which it had become increasingly clear that no one took seriously the possibility that he might have proved Riemann Hypothesis. He was aware of this scepticism. Indeed, in letters from Sarnak and comments from others, it was continually brought home to him that his confidence was not justified. Yet he told me that his manuscript was being typed up as we spoke. 'I shall have it in June,' he said. 'Then I'll work though it in July, checking it, and in August I'll submit it. People don't really work in August, so they may not check it right away, but within a month or so it should be verified.'

'Are you driving in France now, Louis?' I asked, assuming he'd passed his written test and got his licence.

'No, not yet. I'm having to take more lessons. I'm beginning to think I can't drive.'

(14) Taking a critical line

> It is quite a three-pipe problem, and I beg that you won't speak to me for fifty minutes.
>
> Sherlock Holmes, in *The Red-headed League*, by Arthur Conan Doyle

Nowhere is the abstruseness of modern mathematics more starkly evident than when a group of mathematicians assembles to exchange ideas in their own field. Such assemblies are a regular occurrence at the Mathematisches Forschungsinstitut in Oberwolfach, in the Black Forest of Germany, not far from the border with France. The Institute was set up as an international mathematical research centre in 1944 by the *Land* (province) of Baden-Württemberg, which funds a regular series of mathematical meetings. In 2001 there were about forty meetings, most of them on topics at the frontiers of mathematical research. Even those meetings whose titles use ordinary English words convey very little to the non-mathematician:

- Finite Fields: Theory and Applications
- Geometric Rigidity and Hyperbolic Dynamics
- Recent Developments in the Mathematical Theory of Water Waves
- Representations of Finite Groups
- Numerical Methods for Singular Perturbation Problems
- Elliptic and Parabolic Problems of Higher Order
- Ageing and Glassy Systems
- Noncommutative Geometry
- Finite Geometries
- Mathematical Methods in Manufacturing and Logistics
- Dessins d'Enfants

(The last meeting, whose title means 'children's drawings', refers to something described as 'graphs drawn on the Riemann surface'.)

On 16 September 2001, many of the world's experts on the Riemann zeta function began to arrive at the Institute, a modern three-storey building perched on a forested hilltop near the village of Wolfach. The taxis bringing mathematicians from the station passed colourful groups of walkers in knee-britches and hats with flower-bedecked brims. It was the tail-end of the holiday season, and Wolfach is a centre for hill-walkers.

The mathematicians would spend a week here, discussing the theory of the Riemann zeta function and allied functions. The joint chairmen of the conference were Yoichi Motohashi, Matti Jutila and Martin Huxley, each of whom has made a major contribution to the theory of the Riemann zeta function. In addition, Huxley makes regular if minor contributions to literature in the form of limericks about mathematics, a never-ending supply of which flow from his pen (he rarely uses a computer except when compelled to answer e-mail). One of his recent offerings is this:

Montgomery seeks to cajole
Every secret the zeros control.
 Though at present it bothers us,
 The Riemann Hypothesis
Should not be our ultimate goal.

The mathematicians arrived and gathered in the dining area, helping themselves to coffee or wine ('price 1 Deutschmark per glass – please put the money in the wooden box'). Enrico Bombieri chatted in his quiet accented voice to Canadian mathematician John Friedlander; Yoichi Motohashi was in conversation with Henryk Iwaniec; other visitors sat around looking out at the mist-covered hills and the steady drizzle.

Mathematics was not really on anybody's mind that evening. Five days after the New York terror on 11 September, the talk was all about who was coming from the United States and who had been prevented by the ban on flights; what rumours there were about how the Pennsylvania plane had been downed; whether it was possible or wise to make impenetrable cockpits for airliners, and so on.

There's a danger of cliquishness in a conference like this, and the house rules at Oberwolfach try to prevent this by randomizing seating arrangements at the dozen or so dining tables. So at each mealtime the plan of the dining room forms a different random matrix of mathematicians, some of whom are old friends while others have never met.

An air of quiet luxury fills the Institute. Clean, elegant rooms, all provided free along with food; fridges stocked with good-quality German wines at a heavily subsidized price; compulsory coffee and cake between 2.30 and 4 p.m. each day; a billiard table, music room and small gymnasium; and one of the best mathematical libraries in the world. But the heart of an Oberwolfach week is the meetings. The delegates each give a talk about their current work, in a carpeted meeting room with leather chairs, overhead projectors and six blackboards, often entirely covered with formulae and expressions by the end of a talk. Outside the floor-to-ceiling windows is a view of the hills, with barely another building in sight.

The atmosphere was one of hawk-eyed professionalism, as the speakers expanded their ideas with projected slides and chalked formulae. Bombieri's talk was called 'A variational approach to Weil's explicit formula', and suggested an approach to proving the Riemann Hypothesis by proving something else that is equivalent to it.

'In this lecture,' announced Bombieri, 'I want to show how the falsity of the Riemann Hypothesis, which of course it is not…' there is a laugh '…entails unusual consequences for the behaviour of extremals of the Hermitian functional in certain Hilbert spaces.' Occasionally in his talk there was a reminder of what it's all about – the prime numbers. 'There is no prime at infinity – that's clear to me at least,' he said. There are references to 'infinity equals infinity' and 'perturbed and unperturbed kernels'. A small mistake is spotted in a formula with four sets of nested brackets which Bombieri corrected by moving a Greek letter from just inside the second bracket to just inside the third. If a speaker didn't spot his own errors, someone was always quick to point them out.

Then, three minutes over the hour he had been allotted, Bombieri wound up with '…that will do the job and that would imply the Riemann Hypothesis.'

After Bombieri's talk, a Swedish mathematician said, 'Maybe this is where the proof really begins,' and Huxley expressed his reactions in a quickly scribbled limerick:

A prize-winner partly nocturnal
Sends variants of Weil to a journal.
   I'm willing to wager
   This paper is major
(Rank one perturbation of kernel.)

Samuel Patterson gave a talk about something called the circle problem. He wrote a complicated formula on the board, and Bombieri said, 'That can't be right.' He had spotted that there should be a π before the Σ.

At one point in his talk, a Dutch mathematician, Roelof Bruggeman, who had a beard like a hobbit, said, 'It's quite essential that the class number is zero...' which led the audience to shout out in unison, 'One!'

During a talk by Matti Jutila it became clear what a tightly knit group this was, each dependent on some of the others for ingredients in their work. In his talk, Jutila quoted work by Motohashi, Iwaniec, Ivic, Huxley and Conrey, all of them in the room.

Yoichi Motohashi gave his paper on Monday afternoon. It was an attempt to circumvent a process that involves something called Kloosterman sums. The style of his talk and the crispness of his handwritten overheads showed what a fluent and consummate mathematician Motohashi is. From the previous talks, it had already become clear that his own contribution to the field is significant. But with the floor his own, a command of the English language emerged in wordplay that would be elegant from a Briton, let alone a Japanese. The word 'Kloostermania' was written on an overhead. 'I am going to talk about a process of "deKloostermanization",' he said with a twinkle.

One characteristic of the talks at Oberwolfach is that there is always a sense of work in progress, unlike at more conventional meetings where the speakers present cut-and-dried accounts of work in which they have crossed all the t s and dotted all the i s, and don't want to risk being shown up in front of their peers. Here, everyone is very understanding about the rough edges and unfilled gaps in an argument. They'd rather go at twice the speed to discover more quickly whether they are on the right track.

Motohashi said at one stage, 'This is the most exciting point – I haven't done it yet completely, but I'm almost sure.'

And during a talk by John Friedlander, Patterson pointed out something that didn't look right. Friedlander stopped in his tracks, scratched his head, and then said, 'Well, maybe I made a slight error there, but I don't want to get bogged down.' Friedlander filled every one of the six huge blackboards, some of them two or three times over, writing at speed with his left hand in a way which looked very uncomfortable to the right-handers in the audience. 'Can you all see the board?' he asked the audience.

'I can,' said Bombieri.

'Ah, everyone who counts,' said Friedlander.

Even with three-quarters of an hour to speak, Friedlander was in trouble. He wrote faster and faster, truncating here and there to get to his conclusion. At one point he said, 'It's a sum over cuspidal things plus a sum over Eisenstein-type things plus something to do with Klooster-man sums.' In any other mathematical setting a lot more explanation would be required, but most of the people there knew exactly what he meant.

When quoting a result by another mathematician, he wrote 'Weil' but said 'Vile'.

'Do you mean Weil or Weyl?' someone asked. (Modern mathematics has been complicated by the existence of two mathematicians, André Weil and Hermann Weyl, not to mention the recent addition of Andrew Wiles.)

'*Vail* is what he calls himself,' said Bombieri.

'I don't think even he is calling himself anything now,' said Friedlan-der. André Weil died in 1998.

Friedlander continued to write furiously on a board, and when the group suddenly laughed he looked up and asked, 'What did I get wrong?' But they were actually laughing at a mouse that was running across the floor.

Bombieri stood up and held the door open. 'To the library!' he ordered the mouse.

Over breakfast on Tuesday morning, Patterson and Huxley discussed the eccentricities of some of the maths lecturers they had both known when they were at Cambridge. One of the lecturers wore hobnailed shoes and stabbed at the blackboard with the chalk as he wrote, so that the entire lecture was punctuated with rat-tat-tat-tats as he marched up and down, writing and stabbing.

Martin Huxley is a diffident man, hesitant of speech, prone to stare at you while he thinks about what you have been saying – and then fail to reply. He has both the desire and the ability to prove the Riemann Hypothesis. One of the things that becomes clear as you dig deeper is that there's more than one way to state it. 'The Riemann Hypothesis is a precise statement,' Huxley said, 'and in one sense what it means is clear, but what it's connected with, what it implies, where it comes from, can be very unobvious.'

Equivalent statements – mathematical statements which, if they are true, imply the Riemann Hypothesis – can often seem to have little or no connection with the Riemann zeta function. Huxley told me of what are called Farey fractions, which lead to a very simple equivalent statement. They were devised by an Englishman named John Farey, and in a lecture delivered in New York in 1928, G.H. Hardy gave some biographical details of this little-known figure:

> The excursions of amateur mathematicians into mathematics do not usually produce interesting results. I wish to draw your attention for a moment to one very singular exception. Mr John Farey, Sen., who lived in the Napoleonic era, has a notice of twenty lines* in the *Dictionary of National Biography*, where he is described as a geologist… As a geologist Farey is apparently forgotten, and, if that were all there were to say about him, I doubt that he would find his way into the *Dictionary of National Biography* today.
>
> It is really very astonishing that Farey's official biographer should be so completely unaware of his subject's one real title to fame. For, in spite of the *Dictionary of National Biography*, Farey is immortal; his name stands prominently in Dickson's History and in the German encyclopedia of mathematics, and there is no number-theorist who has not heard of 'Farey's Series'.[96]

Farey's series (really a sequence) was obtained by looking at all the fractions between 0 and 1 whose denominator, the number on the bottom, was less than a certain number, say 5, and writing them out in ascending order of size:

$$\frac{0}{5}, \frac{1}{5}, \frac{1}{4}, \frac{1}{3}, \frac{2}{5}, \frac{1}{2}, \frac{3}{5}, \frac{2}{3}, \frac{3}{4}, \frac{4}{5}, \frac{1}{1}, \cdots$$

---

* Actually, the entry on John Farey, Senior in the *Dictionary of National Biography* runs to fifty lines, a rare example of Hardy making an arithmetical mistake.

Farey noticed that every fraction (apart from the first and last) could be obtained by adding together the tops and bottoms of the fractions on either side. So $\frac{3}{5}$, for example, is $(1+2)/(2+3)$ and so on. Hardy said, somewhat cruelly:

> Just once in his life Mr Farey rose above mediocrity, and made an original observation. He did not understand very well what he was doing, and he was too weak a mathematician to prove the quite simple theorem he had discovered. It is evident also that he did not consider his discovery, which is stated in a letter of about half a page,* at all important... He had obviously no idea that this casual letter was the one event of real importance in his life. We may be tempted to think that Farey was very lucky; but a man who has made an observation that has escaped Fermat and Euler deserves any luck that comes his way.[97]

Martin Huxley explained how you can take all the Farey fractions in the interval 0 to 1 with denominators less than a certain number, say 1,000, and write them out in order of increasing denominators (instead of increasing size), so that $\frac{1}{10}$ comes before $\frac{1}{20}$, and so do $\frac{2}{10}, \frac{3}{10}, \frac{7}{10}$, and so on. You then compare this way of writing the numbers with a similar list written out as decimals, in order of increasing size.

'The order in which the fractions come is completely different from their order when written as decimals,' said Huxley, 'and if you try to make this difference quantitative, there's something you can write down about the order which is called a "mean square", and this is as small as possible if the two orders are as different as possible.' Then, as if producing a rabbit from a hat, he produced the Riemann Hypothesis: 'The Riemann Hypothesis says that this mean square is as small as it possibly can be. So in this context the Riemann Hypothesis is saying that the ordering of the fractions as decimals is as different as possible from their ordering as a over b with b increasing.'

In the course of my researches I would quite often come across people who would give me different statements that were equivalent to the Riemann Hypothesis (in other words, if you can prove those statements you have proved the Riemann Hypothesis). There are many such statements, but none of them is simpler than Huxley's.

In addition to statements that are equivalent, there are statements which are true if the Riemann Hypothesis is true. Peter Sarnak

* J. Farey, 'On a curious property of vulgar fractions', *Philos. Mag. J.* 47 (1816), 385–386

explained: 'The Riemann Hypothesis is the central problem and it implies many, many things. One thing that makes it rather unusual in mathematics today is that there must be over five hundred papers – somebody should go and count – which start "Assume the Riemann Hypothesis", and then the conclusion is fantastic. And those [conclusions] would then become theorems. Like this story of this fellow who walked around saying, "I killed nine," and it was nine flies with one swat. With this one solution you would have proven five hundred theorems or more at once.

'The interest of the problem is (a) it's central and (b) it gives you a tool that's extremely powerful. This is one reason we so much want it. It would make life so easy. So when I want to prove something, I might first see if it follows from the Riemann Hypothesis, and if it does then we at least believe it's true.'

Hugh Montgomery described how working on one of the consequences of the Riemann Hypothesis might provide useful information about the hypothesis itself. 'In some ways, exploring the consequences of the Riemann Hypothesis is a way of testing it, so if one were to try to disprove it, one could try to follow the consequences of assuming it and arrive at a contradiction. People in my area spend a lot of time deriving consequences of the Riemann Hypothesis. I have a lot of conditional results myself, pair correlation being one of them, but there are others as well. And so I used to joke that on Mondays, Wednesdays and Fridays I work unconditionally, incrementally, from exactly what is known, trying to push that just a little bit further; and Tuesdays, Thursdays and Saturdays I assume the Riemann Hypothesis and then try to develop a bigger picture. But both types of activity are fruitful in trying to push the subject forward. There are times when one works incrementally and it seems that one isn't making progress, but one is developing tools and getting ideas going, and suddenly the ideas fall into place and things advance substantially.'

So we've got statements that are equivalent to the Riemann Hypothesis, and statements that follow from the Riemann Hypothesis. It turns out that there are also *several different* Riemann Hypotheses. Charles Ryavec still remembers how surprised he was to discover this fact.

'When I got to Ann Arbor,' he said, 'I remember the first year there they had a seminar on the Riemann Hypothesis for *curves*, and I didn't know about that, and I thought, "Jeez, there's another one?" So I went to that talk and that was very interesting. They said there are these similarities, and that really struck me because you've got a Riemann Hypothesis over *here*' – Ryavec waved his hand in the air – 'in a kind of structure which was nothing like this *other* one' – and his hand waved in a different place – 'but then as you develop it, you see this algebra and analysis start fitting together and it gave me a sense then – "Don't try to compartmentalize things too much in mathematics, don't make decisions too fast about how things are supposed to go."'

The 'Riemann Hypothesis for curves over finite fields', as it is called, was proved by André Weil in 1940, while he was in jail for failing to report for duty in the army. Ryavec charmed me with his description of the relationship between the Riemann Hypothesis for curves and the classic Riemann hypothesis, a description which – only for a moment – I thought I understood.

'In the Riemann Hypothesis for curves you've got a blob,' he said, leaning back in his chair and looking towards the ceiling for inspiration. 'Let's just say it's a blob, and there's a function or a map that turns this blob into itself, and you're going to count the fixed points of that action, there's finitely many of them.'

What Ryavec means here is that if you apply a function to a particular curve or surface that acts on each point and transforms it in a systematic way, most points may take new positions on the transformed curve, but a certain number of them, fixed points, could remain in the same place. Such a transformation is called a mapping. You could imagine, for example, a transformation that would take an ellipse and elongate it horizontally to twice its width. In that case there could be two points, at the top and bottom, which didn't change position as a result of the transformation.

'OK, now what you do is you iterate the map,' Ryavec continued, 'so you do it two times, count the fixed points, iterate it three times, count the fixed points, and that's going to give you a sequence of numbers. That sequence of numbers can be encapsulated in a formula, and in the formula are little numbers that have a certain size, and if the certain size is such and such that's the Riemann Hypothesis for curves.'

So, it appears that André Weil's achievement was to show that 'the little numbers were such and such'. But what about the *actual* Riemann Hypothesis?

Ryavec revealed his favoured approach to a proof: 'Now, what's going to probably happen for the real Riemann Hypothesis is there's going to be another blob and there's going to be a function that turns the blob into itself. Something's going to get fixed – it may not be points, it's going to be lines or some geometric figures on the blob, and they're going to have some geometric content, maybe they're lengths, maybe the area, and it's a continuum now, so it's not going to be the map plus its iterates, it's going to be something called a flow.'

I had certainly been able to see how a series of numbers – the numbers of fixed points after each mapping – could demand an explanation, a rule, and get it from Weil. Now I could also see how, if each stage didn't result in fixed numbers but fixed lines or areas, you'd have a much more difficult task to explain how that succession of mathematical objects, called a flow, could be generated. But how would it connect with the Riemann Hypothesis about the prime numbers and the zeros of the Riemann zeta function?

'What that fixed point formula will turn into,' said Ryavec, 'is what we now see as the explicit formulae in prime number theory, so you'll have this beautiful connection. And ultimately these two (the Riemann Hypothesis for curves and the classical Riemann Hypothesis) will be under a common umbrella.'

'And when will that happen?' I asked, since Ryavec made it all sound so straightforward.

'I was once talking to a guy,' he began by way of explanation, 'and I said, "How far do we have to go ahead to see the solution?" And he said, "A hundred years may not be enough. If we go a thousand years maybe it's all forgotten, it's not in the library." But at some point in the future there'll be a book, and it'll contain the Riemann Hypothesis as a special case of some larger theorem, but that may not be the first proof... I don't know. Now someone like me will never get it. It'll be hard to find that blob, how to set it up, how to do the calculations – it's going to be very interesting. A lot of people have worked very hard on it, so you can't really compete with them...'

It seems to be possible to go in all sorts of directions from the Riemann Hypothesis itself – the 'simple' statement that all the zeros of the Riemann zeta function lie on a straight line down the centre of the critical strip at $x = 2$. Its tentacles reach into all sorts of areas of mathematics, and mathematicians in many different fields are trying to find an approach that will work. But progress can be painfully slow.

Martin Huxley has worked in the field for several decades, for much of the time tackling a related problem which all agree is much simpler than the Riemann Hypothesis, and yet that too has provided some pretty tough challenges. It's called the Lindelöf Hypothesis.

Samuel Patterson is another enthusiast for this topic. 'The Lindelöf Hypothesis is an absolutely fascinating hypothesis,' he said. 'It seems that it ought to be provable. It's infinitely easier than the Riemann Hypothesis. Nevertheless, it's way, way beyond anything we can do. I'm sure that to prove the Lindelöf Hypothesis, Martin Huxley would be prepared to sell his soul to the Devil.'

Infinitely easier to prove it may be, but it's not much easier to understand. Like the Riemann Hypothesis itself, it can be stated in a variety of ways. it stems from the work of Ernst Lindelöf, who was active around the beginning of the twentieth century, and it makes a statement about the rate of growth of the Riemann zeta function on the critical line. One version of the Lindelöf Hypothesis is given in Figure 15, comparing the growth of $\zeta(\frac{1}{2}+it)$ to the increase in t.

But the Hypothesis can also be rephrased as a statement about the three-dimensional 'landscape' of the Riemann zeta function that we saw earlier. This is a surface that meanders over a horizontal plane, occasionally dipping to 'sea level' at points where $\zeta(s)$ equals zero. We are particularly interested in the critical strip, one unit wide and stretching off to infinity, where we know all the zeros are to be found. This strip of 'land' has undulations, and the Lindelöf Hypothesis tells us how undulating it is.

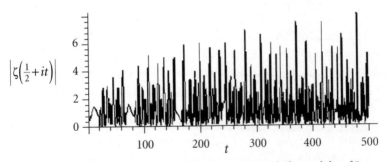

FIGURE 15 One form of the Lindelöf Hypothesis deals with the *modulus* of ζ. (The modulus of a complex number $x+iy$ is the real number $\sqrt{(x^2+y^2)}$.) The Lindelöf Hypothesis says that the modulus of $\zeta(\frac{1}{2}+it)$ grows more slowly than any positive power of t as t increases.

'It is not clear what happens in the critical strip,' Huxley told me. 'Does the graph look like paper, or like corrugated cardboard? By corrugated cardboard I mean that there are fairly regular ridges and grooves which run across the critical strip in the east–west (x) direction. The grooves dip right down to sea level at the values of z which are zeros of the zeta function. Many of these are exactly halfway across the critical strip. The Lindelöf hypothesis says that the grooves are not very high, and the landscape looks more like a piece of paper which is flat for $x > \frac{1}{2}$, and curved upwards for $x < \frac{1}{2}$ as you go left (west). Littlewood proved that the ridges must become low once you go past the nearby zeros. So if all the zeros in the critical strip have $x = \frac{1}{2}$ – the Riemann Hypothesis – then the landscape becomes flat (or at worst gently undulating) for $x > \frac{1}{2}$ – the Lindelöf Hypothesis.'

Again, as with so much in this book, we don't need to be worry about every nuance of Huxley's statement in order to glimpse what he and others have tried to do with the Lindelöf Hypothesis. All we need to know is that there is a mathematical way of measuring the 'smoothness' of the landscape traced out by plotting the Riemann zeta function.

Alexander Ivic is a friend and colleague of Huxley's, and knows a bit himself about the Lindelöf Hypothesis. The measure of 'corrugation' of the zeta landscape is an exponent, a power, in a mathematical expression. To prove that the Lindelöf Hypothesis is true, the exponent must be proved to be zero. Ivic described, in accented English with heavily rolled r s, the painfully slow progress towards that goal.

'The Lindelöf exponent,' he said, 'has decreased from $\frac{1}{6}$ (0.166 666...), a result of Hardy and Littlewood of 1915 or '16 – some time in World War I – to 0.155 something due to Martin Huxley this year, so it took from 1916 up to 2001 – you can count eighty-five years, and it got just that bit smaller. Now, the Lindelöf Hypothesis says that the exponent should be zero, and these improvements from Hardy and Littlewood from $\frac{1}{6}$ to Huxley's latest result – there have been about twenty intermediate results – these improvements do not come from working on more pages, as if you write a paper twenty pages long and then I work half a year more and write a more elaborate paper – no. Each of these tiny improvements on the third or fourth decimal place needed new ideas, fresh views, and this is perhaps the best example to show how unbelievably difficult the Riemann Hypothesis is.'

Samuel Patterson also emphasized the slow progress that had been made on the Lindelöf Hypothesis, even though it is believed to be much easier to prove than the Riemann Hypothesis

'It's proven to be very intractable,' he said. 'The partial results show no sign of converging towards zero, the one-sixth result is about eighty years old, and now it's just a little bit less than that. That has been a result of everything that's been going on in the intervening time, and there's been a vast amount of work done on it. It looks as if it ought to be something that shouldn't cause trouble, but it does.'

Along with Huxley's diffidence went a willingness to talk about his own work on the Riemann Hypothesis, unusual among the mathematicians I had spoken to, with the exception of de Branges. Huxley is prepared to admit that he has an idea, and even to try to describe to me its central elements, presumably because there was little likelihood that I would steal his idea and put the finishing touches to a proof on the train going home.

'The thing I'm interested in,' he said, 'is repeating patterns in two-dimensional hyperbolic space, and there's one particular pattern that's closely associated with number theory. Since I did a lot of geometry at school, I'm very happy with two-dimensional repeating patterns – even if they are in hyperbolic space. Hyperbolic space is a two-dimensional space that curves outwards in the sense that the surface of the Earth curves inwards. It's impossible to visualize.'

Although Huxley says it's impossible to visualize, Figure 16 shows an attempt to do so, in a form devised by the French mathematician Henri Poincaré. There are lots of mathematically interesting things to say about Poincaré's idea, but I shall just mention two of them. This circle is

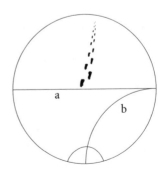

FIGURE 16 The French mathematician Henri Poincaré devised an analogy for a universe that was finite but unbounded. His circular universe, clearly seems finite, but the rules that govern space within it have a number of interesting properties. One is that the nearer objects move towards the circumference, the smaller they become. So someone who walks towards the edge will never reach it. The lines a and b are parallel in the Poincaré universe and meet at infinity. The curve at the bottom is an infinitely long, straight line which meets infinity at both its ends.

a two-dimensional, flat representation of a hyperbolic space. The circumference of the circle represents infinity, so we can have 'parallel' lines, such as *a* and *b*, meeting at infinity. Distances in this model are not fixed. Instead, they are related to the distance of a point from the centre of the circle. The further you go from the centre, the more the distance between points in this space shrinks. If you were walking away from the centre, your steps would get shorter the further you went. As you head for 'infinity', it seems to recede the nearer you approach it. But you don't realize that this is happening because your steps are getting shorter and shorter, since you yourself are getting smaller and smaller, and so is any ruler you take with you to measure things. 'In a sense,' said Huxley, 'the rational numbers themselves form a system of things curving outwards, hyperbolic.'

It's often the case that ideas about this very abstract function, the Riemann zeta function, come down to geometry, either as analogies or sometimes as an interpretation of numbers as coordinates in space, often in spaces of many dimensions. With his hyperbolic space, Huxley believes he's found a way of linking the values in the zeta function to triangles in hyperbolic space. As in any other type of space, a triangle is determined by three points, and the numbers that determine the location of those points in hyperbolic space are derived from the numbers that feature in the Farey-fraction version of the Riemann Hypothesis. Although Huxley believes it's a promising approach, he has no illusions about whether it's *the* idea.

'This is the approach which I hope will do something,' he said, 'and it's something I'd dearly like to make progress in, but one needs more ideas. I reckon for every twenty ideas that I have, one of them turns out to be good. I tend to have my ideas while I'm sitting down for about three hours undisturbed. You can have an idea during a quiet moment at any time of day, and if it's not convenient for family reasons the best thing is to write it down on a piece of paper and not to lose the piece of paper. If it's a really good idea, it'll nag at you until you can work through it.'

Sherlock Holmes described difficult problems as 'three pipe problems'. For Huxley, hard problems in maths are 'three idea problems': 'If a problem needs only one idea, someone's going to solve it. If it needs two ideas, it has to wait for somebody to have both these ideas at the same time, or to have a good memory when they have the second idea. And with the three-idea problem, you need to have three ideas and put them together or you won't get very far. The story behind this classification is that Montgomery solved a problem which I was interested in, using three ideas, all of which I'd had, but at different times. So nowadays I have various pieces of paper simply headed "silly ideas on such and such a subject". And any crazy ideas I have are supposed to be written down, and every now and again I look back over them.'

Abstract delights

> There was a lot of excitement when Connes' ideas came out – partly
> generated by Connes – but it seems that he's hit a brick wall.
>
> Roger Heath-Brown

There is a great gulf between the popular understanding of what mathematics is about and the real nature of the mathematician's preoccupations. This is probably because school maths always starts from the concrete – baths filling, flagpoles casting shadows, weights sliding down rough slopes – and rarely goes beyond that, for most students at least. Even complex numbers, the foothills of abstraction, are dealt with mainly as graphs, which means that learners never think in more than two or three dimensions. The glorious achievements of maths are less accessible than those of almost any other aspect of human culture. When Kurt Gödel received an honorary degree from Harvard for his 1931 paper 'On formally undecidable propositions of *Principia Mathematica* and other related systems', the citation described him as 'discoverer of the most significant truth of this century, incomprehensible to laymen, revolutionary for philosophers and logicians'.[98] In fact, given a little time and a with cold towel around the head, the truth Gödel discovered *is* comprehensible to those laymen (and -women) who choose to work through one of the many popular expositions of it. But there are some mathematical truths that are far more incomprehensible to 'laymen' than Gödel's.

I recently came across a description of a book by a Dr C. Swartz called *Infinite Matrices and the Gliding Hump*, an evocative title, whose blurb has barely a word or expression comprehensible to someone who hasn't done a post-graduate maths course:

> These notes present a theorem on infinite matrices with values in a
> topological group due to P. Antosik and J. Mikusinski. Using the matrix

theorem and classical gliding hump techniques, a number of applications to various topics in functional analysis, measure theory and sequence spaces are given. There are a number of generalizations of the classical Uniform Boundedness Principle given; in particular, using stronger notions of sequential convergence and boundedness due to Antosik and Mikusinski, versions of the Uniform Boundedness Principle and the Banach–Steinhaus Theorem are given which, in contrast to the usual versions, require no completeness or barrelledness assumptions on the domain space. Versions of Nikodym Boundedness and Convergence Theorems of measure theory, the Orlicz–Pettis Theorem on subseries convergence, generalizations of the Schur Lemma on the equivalence of weak and norm convergence in I^1 and the Mazur–Orlicz Theorem on the continuity of separately continuous bilinear mappings are also given. Finally, the matrix theorems are also employed to treat a number of topics in sequence spaces.[99]

It is a measure of the subtlety of the issues raised by the Riemann Hypothesis that a statement which starts with a relationship between the integers and the primes can end up in areas of abstraction that are every bit as rarefied as Dr Swartz's notes on classical gliding hump techniques.

Both in the history of maths, and indeed among the mathematicians I have spoken to, it seems that mathematics gets interesting only when it takes off from the concrete and soars into the realms of abstraction. The question is this: how do mathematicians acquire the taste for such a rarefied diet, and, if society continues to need such people, how is such a taste to be created and nurtured?

It certainly helps if someone with a latent talent for mathematics comes into contact with a good teacher. Charles Ryavec (who was taught by Tom Apostol and taught Brian Conrey) is, by all accounts, a great teacher, and part of his talent lies in the down-to-earth way he tells you things. He is a trim, grey-haired man in his sixties, the only mathematician in a strange institution called the College of Creative Studies, part of the University of Southern California at Santa Barbara, and he has a lot of freedom to teach maths any way he chooses.

'They just give it to me, the freedom,' he said. 'Actually, I'll tell you the truth: most people don't give a damn. We're not on a big budget, it's small potatoes, and I think the university sees this place as window-dressing. They think there's nothing serious going on here, but it's window-dressing and that's it.' He explained how the College came into existence. 'There were people way back when who figured that education was terrible in American universities and what we ought to do is skip kids ahead of these rigmarole courses and get them into exciting stuff. But they couldn't get it done. Universities are very conservative places. Anyway, the Vietnam War came along and the campuses came apart. So this guy, this literature professor here, [Marvin] Mudrick, and a couple of other people who had for a long time had this idea that the educational system was lousy, suddenly said, "Hey, we can get a College of Creative Studies started, this is our dream, this is the kind of idea we've always been thinking of." And the university said, "Fine, do it. Anything to keep the students quiet!" So underneath that idea of keeping the students quiet they got it started. The idea then was that here you would jump students ahead to some kind of vague mix of what they're excited at learning – it's very hard to describe. It wasn't another arts college. There were going to be maths, physics and chemistry as well…'

Ryavec's straightforward speaking style belies the abstract nature of the ideas he is passionate about. He uses words like 'goofy' and 'stuff' a lot, and is very self-deprecating about his own talents. He makes it sound as if he's only in maths because he wasn't much good at anything else.

'At a very early age I fooled around with numbers, but it was on and off again. I had other interests, I love writing a lot, actually I was going to be a novelist for a long time – I am writing a novel. It's my other pastime, I write novels and burn the stuff when I'm done. Actually I've found online a publisher, so I'm going to publish a novel. It's junk, but they accept junk.' He laughed loudly.

What Ryavec's self-deprecating remarks show is a wide-ranging and versatile mind that is capable of coming at maths problems from an unexpected direction, and maybe taking them by surprise with a brilliant insight. He is an unconventional man in an unconventional setting, and he shows passion and tenacity in getting the best out of students who don't fit into the conventional system. The College of Creative Studies is itself an institution that defies categorization. Having read a little about its approach, offering courses in arts and science subjects tailored to the brilliant student, I was expecting some gleaming, light, airy building, designed perhaps to represent the fusion of science and art, music and maths, philosophy and technology. Instead, Ryavec's office is in a building that looks like a morgue, dirty beige in colour, with no obvious entrance, a few small windows and a shabby municipal interior. It had clearly seen better days compared with some of the other gleaming structures on the campus, and Ryavec believed that the College was seen by the current university government as an anachronism or even an embarrassment.

While a more conventional maths faculty tries to attract and process batches of high-school students – some of whom don't even know how to add fractions – and turn them into competent mathematicians, Ryavec has the freedom to pick out high-fliers who may need careful handling but, in his view, have the potential to be much better than competent. As he spoke about his students, Ryavec reminded me of the teacher in the movie *Good Will Hunting* who tames and encourages a mathematical genius in his school. But it's a movie Ryavec loathes.

'I found it a very lousy movie,' he said. 'That movie didn't capture the kind of genius that's going to bubble out of somebody who really finds something. It was all kitschy and goofy, I didn't like the movie at all… I liked the movie *Amadeus*, because I remember when Salieri walks in and you have Mozart there and he plays that sudden improvization – Tom Hulse could do it, you could get this "Ah, this is a genius, this guy." Plus you also saw Mozart up very late at night – he looked awful, and he's writing on this pool table and his wife is saying, "Please come to bed," and he says "No," and Death comes to the door and you really see the kind of obsessional stuff. So I think that kind of genius was done very well in *Amadeus*, but it wasn't in *Good Will Hunting*…'

Like the teacher in the movie, Ryavec keeps an eye out for the bright students who might be rejected by other universities – or even by faculties of his own university – because they wouldn't fit in. 'I've got a student now…who, when I interviewed him before he came here, I knew there was going to be all kinds of trouble. Basically, what I do is if I see him walking here I give him a maths problem through the window.' Ryavec waved an arm towards the window of his cramped office. "Hey, Bob," I'll say, "have you heard of this problem?" And then he'll come by a day later and tell me the solution. That's all we can do. If I were to try to get him to go into a room, he would not do it. He's brilliant. I gave him the other day this problem where you have an integer, like 2,001, and two people play a game – one person takes a divisor of the number and subtracts it from the number, and that's the number he gives the next person. And *he* takes a divisor and subtracts it from the number and gives it back to you, and so the person who can do it and get a 1 wins. So anyway, the student went to lunch and he came back and said, "Yeah, I know how to do that, it's easy. You do this and do that." Now, if I'd given him some function-theory problem, he wouldn't do it…'

Over the years, Ryavec's unconventional approach has nurtured a succession of mathematicians, and a small steady trickle of unusually talented people still find their way to his door, in spite of his pessimism about the way the College is going. 'One kid came in here to the CCS. Before I knew it, he'd disappeared. I thought, "What the hell happened to this guy?" He was over at this advanced course in set theory – it was a really tough graduate course – and so the student came back to me and said, "This professor won't let me into his class," and I said, "Oh, man!" and I went over there and I said, "Look, what's it going to hurt you?" "He's only a freshman," the professor said, "he can't take my course." I said, "Just let him take it, OK?" Finally. So four weeks later I was going down in the elevator with the professor, and he said, "You know, that kid you sent over, he's my grader."* That's how good he was, but all he loved was set theory. Now he's running some big thing for Sun Systems… All the kids I recruited in the first five or six years, almost all went to Silicon Valley. They all live within about a ten-mile radius of each other – some are retired now, millionaires…'

If the Riemann Hypothesis is proved, it will surely be by people like these students of Ryavec's – brilliant, wayward, passionate – who develop a deep knowledge of a small area of maths and think about it morning, noon and night because they find it a more satisfying activity than any of the other pleasures the world has to offer.

For many mathematicians, the sheer buzz they get from doing mathematics outweighs and outlasts almost any other mental or physical pleasure. The earliest mathematicians we know about clearly gained a deep satisfaction from maths. Plutarch describes how Archimedes died during the Roman capture of Syracuse because he was gripped by a mathematical problem:

> A Roman soldier, running upon him with a drawn sword, offered to kill him; and…Archimedes, looking back, earnestly besought him to hold his

---

* He now helps the professor mark the papers of other students.

hand a little while, that he might not leave what he was then at work upon inconclusive and imperfect; but the soldier, nothing moved by his entreaty, instantly killed him.[100]

Ever since, generations of mathematicians have relished intriguing mathematical problems. At the age of eighty-nine, Jack Littlewood was in a nursing home after a bad fall, and his friend Bela Bollobas tried to cheer him up with a maths problem:

> In my desperation I suggested the problem of determining the best constant in Burkholder's weak L_1 inequality. To my immense relief (and amazement), Littlewood became interested in the problem. It seemed that mathematics did help to revive his spirits and he could leave the nursing home a few weeks later. From then on, Littlewood kept up his interest in the weak inequality and worked hard to find suitable constructions to complement an improved upper bound. Unfortunately, we did not have much success so eventually I published the improvement only after Littlewood's death.[101]

'Burkholder's weak L_1 inequality' sounds a most unlikely pick-me-up for an elderly depressed man, but it did the trick in Littlewood's case.

For the rest of us, such pleasures may be difficult to grasp. Apart from, say, meditation or religion, most of us are unused to abstract thinking in our daily lives. This is certainly true of young people at school or college, from whom the next generation of mathematicians will come. Yet somehow each of the mathematicians I met who were working on the Riemann Hypothesis had been able to rise above the mundane landscape of numbers and soar into worlds which not only do not make sense to non-mathematicians but actually seem impossible for the human mind to contemplate. Bertrand Russell had this to say:

> Mathematics, rightly viewed, possesses not only truth, but supreme beauty, a beauty cold and austere, like that of sculpture, without appeal to any part of our weaker nature, without the gorgeous trappings of painting or music, yet sublimely pure, and capable of a stern perfection such as only the greatest art can show. The true spirit of delight, the exaltation, the sense of being more than man, which is the touchstone of the highest excellence, is to be found in mathematics as surely as in poetry. What is best in mathematics deserves not merely to be learned as a task, but to be assimilated as a part of daily thought, and brought again and again before

the mind with ever-renewed encouragement. Real life is, to most men, a long second-best, a perpetual compromise between the real and the possible; but the world of pure reason knows no compromise, no practical limitations, no barrier to the creative activity embodying in splendid edifices the passionate aspiration after the perfect from which all great work springs. Remote from human passions, remote even from the pitiful facts of nature, the generations have gradually created an ordered cosmos, where pure thought can dwell as in its natural home, and where one, at least, of our nobler impulses can escape from the dreary exile of the natural world.[102]

Once again we are faced with a statement about mathematics that is a very long way from *calculations*. The mathematics that Russell describes is akin to philosophy or logic, and much of modern mathematics is like this. Although the starting point for the Riemann Hypothesis is concrete numbers, the methods used in trying to construct a proof are extremely abstract.

The English mathematician J.J. Sylvester, who taught maths to Florence Nightingale, had to be dragged into studying a new and abstract type of algebra by a student, but he didn't regret it:

But for the persistence of a student of this university in urging upon me his desire to study with me the modern algebra I should never have been led into this investigation; and the new facts and principles which I have discovered in regard to it (important facts, I believe), would, so far as I am concerned, have remained still hidden in the womb of time. In vain I represented to this inquisitive student that he would do better to take up some other subject lying less off the beaten track of study, such as the higher parts of the calculus or elliptic functions, or the theory of substitutions, or I wot [know] not what besides. He stuck with perfect respectfulness, but with invincible pertinacity, to his point. He would have the new algebra (Heaven knows where he had heard about it, for it is almost unknown in this continent [America]), that or nothing. I was obliged to yield, and what was the consequence? In trying to throw light upon an obscure explanation in our text-book, my brain took fire, I plunged with re-quickened zeal into a subject which I had for years abandoned, and found food for thoughts which have engaged my attention for a considerable time past, and will probably occupy all my powers of contemplation advantageously for several months to come.[103]

When André Weil had a painful fall as a child, the first thing his sister thought to bring him to comfort him was his algebra book. Later in his life he wrote, 'Every mathematician worthy of the name has experienced, if only rarely, the state of lucid exaltation in which one thought succeeds another as if miraculously… Unlike sexual pleasure, this feeling may last for hours at a time, even for days.'[104]

'If I feel unhappy,' said mathematician Alfréd Rényi, 'I do mathematics to become happy. If I am happy, I do mathematics to keep happy.'[105]

'The joy of suddenly learning a former secret and the joy of suddenly discovering a hitherto unknown truth are the same to me,' wrote Paul Halmos. 'Both have the flash of enlightenment, the almost incredibly enhanced vision, and the ecstasy and euphoria of released tension.'[106]

And according to Thomas Jefferson, '…the science of calculation also is indispensable as far as the extraction of the square and cube roots… algebra as far as the quadratic equation, and the use of logarithms are often of value in ordinary cases: but all beyond these is but a luxury; a delicious luxury indeed; but not be in indulged in by one who is to have a profession to follow for his subsistence.'[107]

For the French mathematician Alain Connes, the pleasure of mathematics can also be accompanied by pain. He speaks of four phases of discovery – concentration, incubation, illumination and verification:

> The process of verification can be very painful: one's terribly *afraid* of being wrong. Of the four phases, it involves the most anxiety, for one never knows if one's intuition is right – a bit as in dreams, where intuition very often proves mistaken. I remember once having taken a month to verify a result: I went over the proof down to the smallest detail, to the point of obsession – when I could have simply entrusted the task of checking the logic of the argument to an electronic calculator. But the moment illumination occurs, it engages the emotions in such a way that it's impossible to remain passive or indifferent. On those rare occasions when I've actually experienced it, I couldn't keep tears from coming to my eyes.[108]

By any standards, Alain Connes is one of the world's cleverest mathematicians. Certainly many of his peers think so. Among the mathematicians working on the Riemann Hypothesis, there were few who would claim to understand Connes's particular specialism, but many who felt that if anyone was going to prove the Riemann Hypothesis, it will be him. For a man who might be considered to have, in Douglas Adams's words, 'a brain the size of a planet', he is remarkably unintimidating.

It was at the 1996 Seattle conference organized by the American Institute of Mathematics that Connes, who, up to that point, had taken no special interest in the topic of the Riemann Hypothesis, was set on a path which some mathematicians believe will lead soon to a proof. Previously, he had been working on the mathematics of quantum field theory and elementary particles, and something familiar occurred to him: 'I was immediately led to the idea that somehow passing from the integers to the primes is very similar to passing from quantum field theory, as we observe it, to the elementary particles, whatever they are.'

Because Connes and a colleague had devised a specific type of algebra to explore this link, he was invited to Seattle to present a paper on it. 'I was invited to this meeting in 1996 and there I gave a talk, but I was very struck by the fact that apart from the work on random matrices not much was going on as far as the Riemann Hypothesis itself was concerned. So then I got really enticed to go on and work on it, and when I came back from the conference, which was in the summer, I pushed the analogy much further. The zeros didn't even show up in my previous work but then, by going further, the zeros came up completely naturally in the sense that one didn't have to define the Riemann zeta function to get the zeros – they were coming from a purely spectral interpretation, and the basic new feature which came up can be explained very simply.'

Nothing Connes says is actually *very* simple – it's the nature of the subject that however 'simply' a mathematician of his degree of specialization tries to explain something, the underlying ideas are so abstract, and any analogies that he uses have to be so mundane, that inevitably you still end up a long way from what the maths is really about.

Over Earl Grey tea in a Paris café called Le Tea Caddy, Connes was expansive and relaxed. Wearing a pale blue sweater and an open-necked shirt, he was animated about mathematics and its importance in his life. He seems to have been passionate about the subject for as long as he can remember.

In 1973, when, at the age of twenty-six, Alain Connes presented his Ph.D. at the *École Normale Supérieure* in Paris, his work on what are called operator algebras was seen as pioneering and described by one observer as 'already a major, stunning breakthrough'.[109] And Connes has since acquired the distinction of creating a new type of geometry that is all his own. As the citation for one of his most recent awards, the Swedish Crafoord Prize, says, he was given the prize 'for his penetrating work on the theory of operator algebras and for having been a founder of the non-commutative geometry'.[110] The Crafoord prize of $500,000, handed to Connes by the King of Sweden in Stockholm in September 2001, is an extraordinary reward for work which, like much pure mathematics, has little or no obvious practical application.

Connes explained to me his view that mathematics is based on a duality between geometry and algebra which, he believes, is reflected in the way the brain works. 'Geometrical perception, which is extremely rich and elaborate, is directly tied up with the visual areas of the brain. Using these areas you can immediately contemplate a picture and perceive the beauty of it. For instance, there is a famous theorem called Morley's Theorem about trisecting the angle of a triangle, and if you draw a picture you will be struck by the beauty of the picture. On the other hand, if you try to write a *proof* by geometrical drawing, in general what will happen is that it will be deceptive. You might have drawn a picture in a certain way which is hiding the fact that you could have drawn it in another way, and then nothing would work. So in general what happens is that the ultimate proof, if you want, is algebraic.'

It's difficult to realize nowadays how unexpected the link between algebra and geometry was to early mathematicians. Even with a minimal grasp of school maths, we nowadays accept without blinking that we can add x^2 to x^3. But early mathematicians would have been shocked at this, as Connes explained.

'It was really a fantastic step,' he said, 'to understand that the square of a number – which is just a geometrical square – and the cube, which is just a geometrical cube – can be added together, even though you would say, "But one has dimension the length squared and the other the length cubed!" and you should never add things which have different dimensions. So algebra is an amazing achievement, and once you have formulated things in algebraic terms then they take a life of their own.'

Connes developed his example of Morley's Theorem, which says that if you trisect all three angles of any triangle – in other words, divide each angle into three equal parts – then the points where the lines of trisection intersect will mark out an equilateral triangle. Figure 17 shows

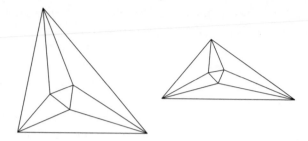

FIGURE 17 Morley's Theorem states that for any triangle, whatever the lengths of the sides, if you trisect each angle (i.e. divide it into three equal angles) then the lines of trisection will all intersect at the corners of an equilateral triangle – one whose sides are of equal length.

how it works. Describing the problem in algebraic terms means treating the triangles as equations with two unknowns, x and y, which allow you to describe uniquely any point or line in the plane. So each corner has its coordinates, x_1, y_1, x_2, y_2 and x_3, y_3, and you can describe the sides and the lines of trisection by equations that have those coordinates as unknowns. Solving these equations using algebra will prove what seems clear from the diagrams – that an equilateral triangle will emerge. Now, if you take the same equations but add another unknown, z, the same mathematical proof will work, and this time it will describe some version of the theorem in three-dimensional space. And by adding more unknowns you can produce a much broader generalization, to four-, five- or six-dimensional space, a concept which is no longer easy to visualize, but makes perfect sense to a mathematician like Connes.

'The same formula will continue to work,' he said. 'The point is that geometry helps tremendously to find a statement, and then you formulate it algebraically. Some people have spoken about the angel of geometry fighting with the devil of algebra, but I believe this is wrong in the sense that what you really have is the duality between the two hemispheres of the brain – and while geometry is located in the visual areas, algebra resides in the linguistic areas. The geographical proximity of words, the way they interact with each other as in poetry, the way they evolve with time, the way they assemble, and so on – that's algebra. What is difficult when one tries to explain what mathematics is about, or decode a mathematical page, is that to a mathematician a bunch of formulas makes sense; it is where the mathematician uses his imagination. He doesn't use his imagination as a science-fiction writer to invent crazy stories or anything like that – no, by no means. He uses his imagination to create mental images from formulas. Gradually, they will begin to make sense in your mind, and even though they will not be as simple as a picture they will have the same type of inner life and dynamics. And then what happens is that the algebra takes over, but it's not that you want to forget the geometry, no – the geometry is still there to back you, but somehow the pictures become more and more [separated] from what we can picture in our ordinary environment.'

For the pure mathematician, moving from two dimensions to three, three to four, and four to many carries with it an idea of a multi-dimensional 'space' in which things exist or happen in the same way as in our familiar three-dimensional space. If I hold the tip of my finger in the air above my desktop, I can describe its position with three numbers: the perpendicular distance from the surface, and the horizontal distances from the two edges of the desk, which are at right angles. I don't know what it would mean to describe it by four figures, or five or six, but a mathematician like Connes can imagine that with ease, and generalize some equations to describe a world in which, just as my three-dimensional finger casts a two-dimensional shadow on the desktop, my three-dimensional finger could be seen as merely a 'shadow' of a four-dimensional finger in another space.

For some time after the AIM Seattle meeting in 1996, there was a feeling among the mathematicians working on the Riemann Hypothesis that with Connes now on the case, there was some hope of a proof. Connes himself was instrumental in organizing a meeting in Vienna in 1998, also funded by the AIM, where he presented his evolving approach to the problem.

Jon Keating did his best to help me understand Connes's approach to the Riemann Hypothesis, with a lot of hand-waving and some very, very broad analogies. 'He's working in a space that's not three-dimensional space, not two-dimensional space, but some really horrible mathematical abstraction. It's an object created from the prime numbers…' Keating then described a tree with a trunk and then branches, and each branch branches off, and those branch off, and branch off infinitely many times. 'You might think of the object we're talking about, the space, as something like this. Well, think of the infinite limit of this object: it's infinitely complicated and convoluted. Alain Connes has a particle moving in a complicated space, and the space is created from the prime numbers, and again in the loosest possible way, the lengths of the branches of the tree are related to the logs of the primes. And that's how the primes come in.'

I quite like the way the branches of the tree are related to logs, but in spite of Keating's valiant efforts, the abstractions of Connes's mathematics eluded even his explanatory powers. This is a measure of the extremity of the frontiers of mathematics at which Connes is working.

By 2001, Connes was at pains to point out that he wasn't claiming to be near a proof, which is just as well, because at about the same time many mathematicians were saying that, whether or not he had been near a proof, he certainly wasn't now. Andrew Granville was one of those.

'This stuff that Alain Connes did – for people who really understood it – it was definitely very exciting,' said Granville. 'Maybe the guy had a way in from a completely ridiculous angle. It was so different from what a lot of people were thinking. I personally am not capable of judging Connes's work, but my impression from various people who are studying it is that it seems that, though he's had some beautiful ideas, it doesn't seem like he's got the depth of ideas that are going to turn the problem around. He's got some beautiful thoughts, he's got new observations, he's a great mathematician and he was optimistic that certain things would work out, but he's not a great expert on the Riemann Hypothesis, and when the experts started looking at [Connes's work] they realized that maybe he's dressing up in another language a well-known difficulty in the usual language. But that's not to say he doesn't deserve credit for fantastically interesting stuff. In some sense, it's got to be valuable in one way or another in the long run.'

What's unusual about Connes is that although he's been a mathematician all his life, he's only recently been seized by the Riemann Hypothesis bug. For many of the others working seriously on the Riemann zeta function – de Branges, Iwaniec, Bombieri, Conrey, Heath-Brown – it's been the only topic they've really ever wanted to work on. But the late converts are often the most passionate ones.

'It is probably the most basic problem in mathematics, in the sense that it is the intertwining of addition with multiplication,' Connes said. 'It's a gaping hole in our understanding, because until we really understand it we cannot say that we understand the line. Even the line itself is still extraordinarily mysterious.'

'The line' for Connes is not just the line of real numbers. He works with a line that has whole sets of other numbers, called p-adic numbers, each set containing an infinity of numbers based on a different prime number, hence the p in p-adic. One indication of the unexpected nature of the p-adic numbers is in the relationships between the numbers on the p-adic line. Their proximity to one another is related to the power of the prime P that divides their difference. In other words, two p-adic numbers, x and y, are near to each other if their difference, $x - y$, is divisible by a very large power of the prime P. All the different p-adic numbers form what's called an adele.

'It is a basic primitive question about the adelic line which we don't understand,' Connes said. 'It is a question about the way addition is fitting with multiplication. It is a basic problem – we cannot dismiss it by saying that is not interesting. With a problem like "Are there infinitely many powers which are adjacent? Like 8 and 9 (2^3 and 3^2)?", well, you can dismiss these problems very easily as being uninteresting. But a problem like the Riemann Hypothesis cannot be dismissed as uninteresting. It is really one of the most basic questions.'

After Seattle, Connes became intrigued by the work of Berry and Keating, especially by the possibility of realizing the Riemann zeros as a particular spectrum. But there was a problem with this. Berry and Keating kept finding that a particular sign that should be positive if their guess was right is actually negative.

'Whatever you try to do to fiddle, to correct it and so on, it doesn't work, it never works,' said Connes, 'but the idea which came up in my work is the following: that when you're trying to find a spectral interpretation you look for an *emission* spectrum, but physics has more in store. Let me be more specific: when you heat some pure chemical substance, you can let the radiation that comes out of it go through a prism and then you see spectral lines. You see a few bright lines on a dark background. But it turns out that in physics there are other spectra, such as the light coming from distant stars, which are called *absorption* spectra and which are exactly the opposite. What you receive is white light with all frequencies present – except that there are a few dark lines, and these dark lines signal the fact that when the radiation is going through the outer atmosphere of the star there are some chemicals which absorb it.

And in my own realization of the zeros as a spectrum, they appear not as an emission spectrum but as an absorption spectrum, and this exactly accounts for the minus sign which was present in the Berry–Keating approach.'

Michael Berry is intrigued but not convinced by Connes's new idea. 'If you talk to him he will talk about an absorption spectrum. Now what he means by that is that he thinks that it might be the case that the Riemann zeros are not energy levels of a system but rather the following: you've got a continuum of levels, but there are certain levels where it's impossible to make states, so there would be anti-levels, gaps, which represent the zeros, so he thinks. That's what he means when he says absorption spectrum. Now, we think this is a matter of formalism and that everything you can write as an absorption you can write as an emission spectrum if you manipulate things the right way. We're discussing this with Connes at the moment, but that's his idea – it may be right or it may not be.'

Gripped by his new approach to the Riemann Hypothesis, Connes spent two or three years after Seattle working on what's called a trace formula, which he thought could lead directly to a proof of the Riemann Hypothesis. 'I refined the formula and checked a lot of examples and I found that it works beautifully when you only consider finitely many primes but it doesn't seem to be so easy to justify in any way when you consider infinitely many primes. I believe I have found a very nice framework but this framework is still awaiting the main actor. So there is the stage – it is perfectly well arranged and so on – but we are still expecting the heroine to come and complete it. And, of course, you know, until you actually do that there is no way to estimate the distance to target.'

There's a disarming quality about discussing advanced mathematics with Connes. You feel that he's trying so hard to make it comprehensible that you're being unreasonable by not always grasping the essence of what he's saying. He even *says* how simple it all is: '...the starting point is extremely simple...a trace formula is something which is quite simple to explain...mathematically, it is rather simple to explain...'

But if most of us find mathematical formulae inscrutable, Connes has an intriguing theory why this is and why it might not always be so. He compared reading mathematical expressions with reading music. 'It could be that we are in the situation that people were in when nobody could "hear" music when they looked at a score. A few people *are* able to, of course. It's like if you take a conductor, he is able, when he sees a score, to make sense out of it without playing, but most people are not able to do that. With mathematics it's a little bit as if we were in a very uncultured society where music is still at an incredibly early stage of development, and where the effect that music has on people still varies tremendously from one person to another, and where it is not yet perceived as a language. It's clear that when you look at a score by Chopin for example, a very precise one, there is so much care taken in the writing, there is so much thought about the interplay, that it is not very far from an extremely elaborate piece of mathematics. There are a very few select people who can think abstractly of music at a very high level, and there are very few people who can think mathematically in the same way…'

16 Discovered or invented?

> I don't deal with philosophical questions – they are too unnerving.
>
> Alexander Ivic

After fifteen chapters of a book on the Riemann Hypothesis, I have to break some bad news to you. You know almost nothing about the Riemann Hypothesis compared with what there is to know. Obviously, that's because it's impossible to convey to you what you don't know, and the reason you don't know is that it's impossible to convey. This is true of most modern maths, and it's not just because these things are difficult in the sense that a very hard sum is difficult. They are difficult because philosophically the mathematician's mind works on a different plane, a plane that rarely intersects with the thinking of the rest of us.

For me, nothing brings home this difference more than mathematical jokes. Here's one to think about. A topologist walks into a bar and orders a drink. The bartender, being a number theorist, says, 'I'm sorry, but we don't serve topologists here.' The disgruntled topologist walks outside, but then gets an idea and performs Dehn surgery upon herself. She walks into the bar, and the bartender, who does not recognize her since she is now a different manifold, serves her a drink. However, the bartender thinks she looks familiar, or at least locally similar, and asks, 'Aren't you that topologist that just came in here?' To which she responds, 'No, I'm a frayed knot.'*

I suppose it's possible that the topologists' mathematical joke is no more inscrutable than any other joke that uses technical terms. There are probably doctors' jokes based on obscure diseases, medievalists'

---

* If you want to know what Dehn surgery is – if you *really* want to know – it is defined as 'The operation of drilling a tubular neighborhood of a knot K in S^3 and then gluing in a solid torus so that its meridian curve goes to a (p, q)-curve on the torus boundary of the knot exterior. Every compact connected 3-manifold comes from Dehn surgery on a link in S^3.'

jokes based on rare texts and pilots' jokes based on aeroplane components. But let's try another one:

2 plus 2 equals 5 for sufficiently large values of 2.

This is subtler. At least with the 'frayed knot' punchline you can make
a guess at why it's funny to topologists. Here, we need to be steeped in
maths terminology to the extent that we know that mathematicians
often discuss the behaviour of mathematical functions of x 'for
sufficiently large values of x', the point being that x is variable and so
can have large or small values. Whereas 2 is not variable, it's just 2.

Then there are jokes which are a little deeper. Littlewood wrote about
his favourite joke, and added a remark which I find funnier than the
joke:

SCHOOLMASTER: 'Suppose x is the number of sheep in the problem.'
PUPIL: 'But, Sir, suppose x is *not* the number of sheep.'
(I asked Prof. Wittgenstein was this not a profound philosophical joke,
and he said it was.)' [111]

André Weil told a true story with similar philosophical overtones, about
the time he spent teaching troops in America who were waiting to travel
to the war in Europe:

I wound up having to spend fourteen hours a week spoonfeeding these
poor boys the elements of algebra and of analytic geometry. They showed
up in class in uniform, marshaled by a non-commissioned officer. One
day, one of them had a question: 'I don't understand what x is.' The question was far more profound than he suspected, but I did not attempt to
explain why. [112]

Profundity often lies at the heart of mathematical humour, more so
than with other types of joke, and what is profound about mathematics
often seems humorous to outsiders. Lewis Carroll – who was a mathematician – knew this, and modern maths has creations that might well
have been plucked from *Alice's Adventures in Mathsland*, if there were
such a book. We've already come across the gliding hump, for example.
Then there are large cardinals, the hairy ball theorem and different
orders of infinity – some bigger than others. There's the situation where
100 per cent of a collection of items isn't necessarily all of them – in fact,
you may have 100 per cent of them but there are still an infinity of them

left over. And then there's Gabriel's horn, a three-dimensional shape with infinite surface area but finite volume. And so on and so on.

What *is* this world we've been exploring? Could it contain anything the human mind chooses to devise, any oddball concept that a mathematician feels like describing? Or does it have limits, imposed by the very nature of mathematics itself? Fantasy or reality? Invented or discovered?

There's one metaphor that comes up time and again as mathematicians attempt to describe their work: the metaphor of landscape. The Canadian humorist Stephen Leacock chose this metaphor to characterize the difficulties he found when learning mathematics:

> How can you shorten the subject? That stern struggle with the multiplication table, for many people not yet ended in victory, how can you make it less? Square root, as obdurate as a hardwood stump in a pasture – nothing but years of effort can extract it. You can't hurry the process. Or pass from arithmetic to algebra; you can't shoulder your way past quadratic equations or ripple through the binomial theorem. Instead, the other way; your feet are impeded in the tangled growth, your pace slackens, you sink and fall somewhere near the binomial theorem with the calculus in sight on the horizon. So died, for each of us, still bravely fighting, our mathematical training; except for a set of people called 'mathematicians' – born so, like crooks.[113]

But mathematicians, crooks or not, are made of hardier stuff than Leacock. It's clear that the landscape of mathematical exploration has a vivid and challenging reality to them. As they describe their past and present lines of attack on their own particular fields of study, it is as if they see in their mind's eye a landscape with jungles and marshes, paths and rivers, cliffs, foothills and mountains:

> The process of mulling over a mathematical problem displays a striking similarity to that of surveying a cliff before the ascent; of visualizing and comparing alternate routes, from the big lines of ridges, ledges and chimneys down to the details of toe and finger holds, and then weighing possibilities of what might be encountered beyond the visible; all in perfectly focused concentration, projecting ahead, extrapolating, performing so-called *Gedankenexperimente* (thought experiments) and sensing them throughout one's bones and muscles.[114]

Whymper made seven efforts before he climbed the Matterhorn in the 1860s and even then it cost the lives of four of his party. Now, however, any tourist can be hauled up for a small cost, and perhaps does not appreciate the difficulty of the original ascent. So in mathematics, it may be found hard to realize the great initial difficulty of making a little step which now seems so natural and obvious.[115]

If we compare a mathematical problem with an immense rock, whose interior we wish to penetrate, then the work of the Greek mathematicians appears to us like that of a robust stonecutter, who, with indefatigable perseverance, attempts to demolish the rock gradually from the outside by means of hammer and chisel; but the modern mathematician resembles an expert miner, who first constructs a few passages through the rock and then explodes it with a single blast, bringing to light its inner treasures.[116]

In the face of so much gritty reality it may come as a surprise that it's possible to question the very existence of mathematics as anything other than a creation of the human mind. It's the sort of thing philosophers do, perhaps for the sheer hell of it. I spoke to Adrian Moore, a philosopher at St Hugh's College, Oxford, about the ideas of many, perhaps most mathematicians, who are realists and believe that maths is 'out there' in some sense rather than 'in here'.

'One of the biggest worries anti-realists have about realism', he explained, 'is the question of how we have access to these mathematical entities. OK, no realist is going to believe that numbers and sets and suchlike exist in exactly the same way as physical objects. On the other hand, if you're going to talk in terms of mathematics existing somewhere "out there" at all, at the end of the day there must be some sort of analogue of perception. We have a reasonably good understanding of what it is that gives us access to physical objects, and indeed it is itself something that can be the object of scientific investigation. You can do physiology and look at how we come to know about physical objects. And the worry is that there's no convincing equivalent story that you could tell about mathematical objects. That's one of the things that pushes people in the direction of anti-realism – they are inclined to say, "No, really the only thing at the end of the day that we can see clearly and can clearly make sense of is mathematical practice itself." And it's best understood in its own terms rather than as giving us this mysterious kind of access to something independent.'

But Samuel Patterson will have none of this. He just *knows* that mathematics is real. 'I feel very much in studying the theory of numbers that I am studying something that is already there. The analytic theory of numbers in this sense had its beginnings about two hundred years ago when people started compiling large tables of prime numbers, and they discovered that although the prime numbers themselves are very regular, it's very difficult to predict whether a given number is going to be prime or not. The [whole] numbers behave very smoothly, very elegantly, and you can say that the probability that a number is a prime number is more or less the inverse of the number of digits it has – this is one way of phrasing the Prime Number Theorem. And this was an experimental discovery in this sense and it was then a challenge to the community of mathematicians to explain this phenomenon.'

The notion of 'experimental discovery' in mathematics is an intriguing one. It may seem odd to talk of experimenting with something you can't touch or manipulate in any way, to test your theory by experiment. But many branches of science have theories based on data that the scientists cannot manipulate. Darwin's theory of evolution by natural selection is founded on a combination of natural observation and fossil evidence from the past. The timescales are so huge that you couldn't test the creative power of evolution on any human timescale with most creatures. Similarly, astronomers and cosmologists gather existing data and dig around in them to construct theories that they can then test against other data they may gather in the future. That's how Patterson sees the task of exploring and explaining the phenomena of the Riemann zeta function.

'It was almost in the same way as you might say you're trying to explain something physical, the orbits of the planets or some other phenomena in the real world,' he said. 'It was just before 1900 that the Prime Number Theorem was eventually proved. In the course of that, it turned out that the distribution of the prime numbers had a much more intricate structure than one had imagined before this. And this intricate structure was what was discovered by Riemann and is encapsulated essentially by the so-called zeros of the Riemann zeta function. So this is a very characteristic development – you aim at a particular problem and you discover that there is much more to it than met the eye. All problems are not like that – some problems turn out to be silly, but with certain problems – and in the case of the distribution of prime numbers it turns out that there is much more structure than one would have imagined at first sight.'

Matti Jutila, too, chooses an astronomical metaphor. 'I sometimes have the feeling that number theory is comparable with the empirical sciences – we are dealing with a phenomenon. In physics or natural sciences, the topic of research is something which is given from outside: it is there, but in mathematics the principal object of study is an abstraction. Mathematics begins from nothing, and there are axioms and definitions, and then we prove theorems. One may think that these are pure abstractions, but still, although I am not a logician, I have a feeling that the number system is comparable with the universe that the astronomer is studying when he finds all sorts of phenomena in the cosmos. The number system is something like a cosmos.'

Yoichi Motohashi, for whom prime numbers are as solid as a metal lampshade, finds, when pressed, that the question of the reality of mathematics is not so simple. 'Whether mathematics is a creature of mankind, or has physical existence and is something independent from human beings, is a very hard problem. If mankind were to disappear completely, what is mathematics? Will it still remain as mathematics? Maybe the integers are the creation of the human mind.'

He then told me a story about his daughter Haruko's approach to whole numbers when she was a small child. 'She was not very good at mathematics or arithmetic. One day I asked her the following question: "There are three cakes on a plate, and I add one cake, and you take one cake. How many are remaining?" And she said, "Two." I said "What? What did you say?" I was a bit upset. I asked again, "OK, three cakes, add one and you take one – how many?" And she said, "Two." And I kept asking her "Why?" She started crying, and I was very sorry. Then she went to her mother and explained why she thought so. Japanese cakes are sticky, and if she picks up one, the other will always come with it. First you have four, and if you pick up one the other one sticks to it and two remain. So she was logically correct.'

It has always seemed to me that what would sort the realists from the anti-realists is the question of whether extraterrestrials would recognize the unusual nature of prime numbers and, by extension, develop number theory and perhaps even prove the Riemann Hypothesis. This is certainly the belief of space scientists who transmit radio messages from Earth in the hope that they will be intercepted by some galactic

FIGURE 18 This image was encoded in a string of 1,679 0s and 1s and transmitted by radio waves into space. The hope is that any extraterrestrials who intercept this message will realize that 1,769 is 73 multiplied by 23, two prime numbers, and will plot the 1s and 0s on a 23 × 73 grid. If they do so, they will see simplified diagrams of a number of concepts that are important to Earth people, from atomic structure to DNA, the Earth's population and our position in the solar system.

civilization and recognized as sense rather than nonsense. One message transmitted in 1974 was composed of 1,679 bits of information. The assumption was that any extraterrestrial at least as bright as us would recognize that this number could only be factorized in one way, 73 times 23, and would therefore see that it should be arranged as a grid to produce the picture shown in Figure 18.

'The image contained in the message reveals many things about ourselves and our home,' say the scientists who devised it, in a victory of hope over expectation. 'It shows our chemical make-up and height as well as the population of the Earth. It also shows our Solar System, and the Arecibo telescope which transmitted the message.' The pair of curved lines in the middle, for example, represents the DNA double helix.

It may seem obvious that beings everywhere have to count, and therefore that it wouldn't take them long to arrive at the primeness of numbers. But anti-realist philosophers such as Adrian Moore can always wriggle their way out of such obviousness. He used as an example the human appreciation of music.

'We might come across extraterrestrials,' he said, 'who just couldn't see what was going on with a kind of music that we enjoyed, for example, if you played Beethoven to them and it's just noise. Suppose they have the same senses, and they can hear it but it just doesn't make any sense for them. We could then find that they go to concerts and listen to stuff that to us is a cacophony.'

What I like about philosophers is the way they pursue their analogies. Moore's example set me thinking about the likelihood that extra-terrestrials would (a) go to one of our concerts and (b) use valuable accommodation space on their ships to bring their own orchestras with them to play at specially organized concerts of their own. But Moore then explained the relevance of this bizarre picture.

'With music, that wouldn't be particularly surprising – we can get our heads round that. I think that an anti-realist would have to be prepared to say it could be like that with mathematics. That the extraterrestrials may use all sorts of symbol manipulation which serves certain purposes for them – perhaps even helps them build bridges, who knows? – but when they try to explain what's going on to us, or we try to explain our mathematics to them, neither of us gets it.

'Take the multiplication of two minuses. Suppose they said, "I just can't see what you're on about – you multiply minus three by minus three and you get minus six, not *plus* six," and you could imagine that eventually it might just flounder. And if in fact Martians as a whole had that sense of consternation but were in other ways obviously very sophisticated, if you are an anti-realist you'd accept it and say, "Well, OK, there are two different ways of doing maths."'

To a strong realist such as Alain Connes, such ideas are heresy. In fact, he believes that mathematics is *more* real than what we think of as solid external reality, that instead of mathematics being embedded in the physical world, the physical world is embedded in mathematics.

'It is quite amazing to realize that our simple-minded materialistic conception of external reality is really built on quicksand,' he said, 'and that after a while the only real thing you can cling to is much more abstract. Just let me give you a concrete example of that – if you take one individual, most of his cells are actually replaced totally over a period of several years, so what is he? Is he a collection of cells? Certainly not, because precisely these cells are replaced. But what he is is something quite different – it's a scheme. The only thing that is pertinent is the scheme, the organization. Quantum mechanics is extremely striking in that respect, because it makes it clear that even if you try to cling to external matter as being reality, you will find that as soon as you go to the sufficiently small, then precisely because of quantum mechanics you will come across inconsistencies, so you can't rely on that as being the ultimate reality.'

Connes's passionate realism is born from experience of the deep 'realities' of mathematics. It surprised me, therefore, to discover that Henryk Iwaniec, equally experienced in the profundities of number theory, felt differently

'I think it's invented,' he said. 'I know that I say, "We have these beautiful things," and "We are trapped here." or "We are trapped from every other direction," or I say, "You know there must be some forces above that have this under control." But this is just a joke, I think – we invented everything.'

I then asked him the question about alien intelligences – would they have different mathematics from ours?

'Yes,' Iwaniec said firmly.

'They wouldn't have prime numbers?' I asked.

'Ah, that far?' He hesitated and thought for a moment. 'Probably not. I'm not good at science fiction, I don't read much of it. But possibly so.'

But there are some mathematicians working on the Riemann Hypothesis for whom these arguments would just get in the way of the fun they're having. 'I never think about such things, never,' said Peter Sarnak. 'We know what we're after, we're working people. When you're out there in the front line, you can't start worrying about whether numbers are real or what they mean to you.' Then, Sarnak did 'think about such things' for a moment. 'I think in many proofs you may say we discover them, but what we're trying to prove is kind of God-given.'

Alexander Ivic, the tall, gravelly voiced Serb, said merely, 'This is a philosophical question, and to be frank I don't deal with philosophical questions – they are too unnerving, so to speak.'

For many who do deal with these questions, the most powerful argument for the reality of mathematics is the fact that it can lead to correct descriptions of the physical world and practical applications of science. It has to be said, however, that this is not the prime motive for most mathematicians, as this story about Euclid suggests:

> Someone who had begun to read geometry with Euclid, when he had learned the first proposition, asked Euclid, 'But what shall I get by learning these things?' whereupon Euclid called his slave and said, 'Give him three-pence, since he must make gain out of what he learns.' [117]

Archimedes, too, was scornful of the need some people felt to justify mathematics on the grounds of its practical applications:

Despite the great success of Archimedes' military engineering inventions, Plutarch says that 'He would not deign to leave behind him any commentary or writing on such subjects; but, repudiating as sordid and ignoble the whole trade of engineering, and every sort of art that lends itself to mere use and profit, he placed his whole affection and ambition in those purer speculations where there can be no reference to the vulgar needs of life.'[118]

Nevertheless, thanks to what has been famously called 'the unreasonable effectiveness of mathematics in the natural sciences',[119] whatever mathematical topic mathematicians pursue for the sheer pleasure of it could, sooner or later, turn out to give some physical description of the world or have some practical application. When Einstein was seeking an explanation for some of the observations of early twentieth-century physics, he turned to an entirely 'useless' area of mathematics devised a few years earlier. In 1907, Hermann Minkowski, as an exercise in abstraction, devised a theory that linked space and time in a 'spacetime continuum'. Then, when Einstein needed a mathematical theory that fitted the observations, he saw that Minkowski's geometry was tailor-made for him. It's arguable that the theory of relativity wouldn't have happened if Minkowski's ideas had not been available.

The Riemann Hypothesis has been around for longer than Minkowki's theories of spacetime, and hasn't yet found an Einstein to turn it into nuclear power stations and atomic bombs. Yet it would be surprising if the research it has stimulated and the new fields of mathematics it has spawned didn't lead to future developments in how we understand the world.

Andrew Granville gave me an example of how number theory, derived purely from a fascination with the integers and how they behave, now has unexpected practical applications. He pointed at the recorder I was using to record my interview with him.

'That recording device has got number theory in it. The reason it's fairly accurate in reproducing sound is that it has error-correcting codes in it which are an application of elementary combinatorial number theory. It's an idea from ten or fifteen years ago that first got put into CDs when they were developed and is now in all electronic recording devices. In a year's time, if you pick up your cellphone and phone home you'll probably have a cellphone that will encrypt your voice in real time and the encryption of your voice will be done by number-theory methods.'

Any encryption system has to balance two opposing pressures – to make decrypting as easy as possible for the legitimate recipient and as difficult as possible for anyone else. A system called RSA cryptography (the initials are those of the people who devised it) is based on very large numbers which have been created by multiplying together two very large primes. The code can only be broken if you know the primes, and factorizing very large numbers is very time consuming.

What mathematicians have been trying to do since ever this system was published in 1977 is to devise faster ways of factorizing larger and larger numbers. In 1994, for example, the best method of factorization was tested on a number with 129 decimal digits. The entire factorization took the equivalent of 8,000 years of computer time at a million instructions per second. Much of the work was done using 1,600 computers working in parallel, but even then it showed that a 129-digit number would take a long time for a potential spy to factorize, by which time the information would probably be out of date. But computers got much faster, and a number of 129 digits was soon beginning to look too small. A test was done in 1999 on a number with 155 digits (100,000,000,000, 000,000,000,000,000 times larger than the number with 129), and it took the equivalent of over thirty-five years of computer time to factorize, which, because the computers were working in parallel, amounted to 'only' just under four months of real time.

But as computers get faster and have more memory, cryptographers have to devise ever better methods of encryption, and these will rely upon the discovery of more and more sophisticated mathematical methods, some of them derived from a proof of the Riemann Hypothesis

'What's become an interesting mathematical issue is coming up with a cryptographic scheme which can run amazingly fast,' Granville said. 'The mathematicians who are working on it are trying to reinvent the number theory behind cryptography. My involvement is for academic interest, but I believe that the money involved in having the best protocol is between a billion and ten billion dollars a year. So you're talking enormous sums of money, but what's nice about it in some sense is that it gives you interesting maths questions to think through. Now, I've no objection to working on a problem that's of significant interest for the real world. But I don't think for anybody that is their primary interest – they kind of fool around with it, then when they discover something, they say, "Hey, maybe I can get rich!" Then they try to find venture capitalists who'll make them rich.'

Without a proof or disproof of the Riemann hypothesis neither the Clay Prize nor a slice of a billion-dollars-a-year pie will go to any of the mathematicians working on it. At the beginning of 2002, only one mathematician was saying publicly that he thought he was within sight of a proof. And we can guess who that was.

(17) 'What's it all about?'

It would be a tragedy if it just needed a trick to prove it.

Alain Connes

T.S. Eliot's widow, Valerie Eliot, once wrote this letter to the London *Times*:

> My husband, T.S. Eliot, loved to recount how late one evening he stopped a taxi. As he got in the driver said, 'You're T.S. Eliot.' When asked how he knew, he replied, 'Ah, I've got an eye for a celebrity. Only the other evening I picked up Bertrand Russell, and I said to him, "Well, Lord Russell, what's it all about?" and, do you know, he couldn't tell me.'

As applied to the Riemann Hypothesis, 'What's it all about?' is a question I've tried to answer in a number of different ways, and a taxi driver reading this book may still feel that I haven't told him (or her). But then I couldn't teach anyone to drive a car solely by writing a book, though I could perhaps give them some idea of how an engine works, how much fun it is driving fast through the countryside on a sunny day, and the number of interesting places you could visit if you *could* drive a car.

So, it's worth recapping some of the steps that have led to this famous statement in mathematics, and trying to clarify where mathematicians stand in their attempts to prove it.

The Riemann Hypothesis matters because, if it is true, it proves that there is a rule for generating the prime numbers, the building blocks of all the other numbers. At the moment, we cannot prove that such a rule operates. If you look at where the primes fall in the infinitely long list of whole numbers, there seems to be no pattern – the distribution looks random. But Bernhard Riemann identified a mathematical function, now called the Riemann zeta function, that was intimately related to the infinite set of prime numbers and that generated another infinite set of numbers, called the zeros of the function. If those zeros all behave in the way Riemann believed, they will allow us to describe exactly how the prime numbers are distributed, as far down the road to infinity as we care to go.

As Gauss discovered, if you plot the number of primes below a given number on a graph in a certain way, it increases fairly steadily. But the actual value fluctuates erratically around that steadily increasing line. If you remove the effect of the increase and just look at the fluctuations, you will find that they are like the massed sound of an orchestra, with each of the zeros contributing an element – a pure tone – to the complex wave. The more zeros ('tones') you add, the closer the total graph of the zeros is linked to the fluctuations of the primes. With an infinite number of zeros added in, the curve becomes locked in with the graph of all the primes up to infinity. As Andrew Odlyzko said to me, 'All of the zeros together determine all the primes, and vice versa.'

The Riemann zeta function is a mathematical expression involving powers of prime numbers and one unknown quantity, usually denoted by s. If you substitute certain numbers for s, the whole expression will boil down to zero (just as $s^2 - 4$ boils down to zero if you substitute 2 for s). The values of s that make the Riemann zeta function equal to zero are complex numbers – numbers with two components, a and b. The first is called the real part of s; the second is the imaginary part, because it is multiplied by i, an imaginary number that stands for the square root of minus 1. Although you can substitute any imaginable complex number into the Riemann zeta function, with any values of a and b, the only ones that have ever been found to make the Riemann zeta function equal to zero are those in which a is $\frac{1}{2}$. Since we can represent numbers of the form $a + ib$ as points on a graph, a units from one axis and b units from the other, all the numbers with a equal to $\frac{1}{2}$ are said to lie on a straight line half a unit from one axis and stretching off into infinity.

An analogy may help us. A Mr Bernhard Riemann has a bank account at the Zeta Bank. As with most people, his bank account is sometimes in credit, sometimes overdrawn and occasionally neither – when it has nothing in it at all. Again like most people, Bernhard has an income and he has expenses. The balance in his account is determined by with-drawals – cheques he writes or direct debits he has set up – and deposits – salary from his employers, plus occasional gifts of money from a maiden aunt or winnings from occasional bets on the horses.

Over the years, Mr Riemann's transactions have generated a huge pile of bank statements which chart the hour-by-hour fluctuations of his account. Zeta Bank is very meticulous about these things, and wants to be absolutely sure of the balance of Riemann's account at every moment. An interested observer, who happens to be a tax inspector called Titchmarsh, is flicking through the bank statements idly one day when he notices a pattern in the occasions on which Riemann's Zeta account has a zero balance – when it is neither in credit nor overdrawn. This seems to happen only at 12 noon on some Wednesdays. Not all Wednesdays, not even Wednesdays at specific intervals. It's just that the only times he ever finds Riemann's Zeta account at zero is on a Wednesday at noon.

This all seems pretty suspicious (but then tax inspectors have a low threshold of suspicion – they don't trust anyone). So Titchmarsh gets hold of all the statements, many years' worth, and tests his hypothesis. Sure enough, the only Riemann Zeta zeros he can find are at noon on Wednesdays. But as he looks more closely at the pattern of Riemann's income and expenditure, he discovers that there *is* a rule that says that the zeros can *only* occur on Wednesdays. The pattern of timings of Riemann's salary cheques and direct debits, combined with Riemann's inactive lifestyle – he stays in bed most days and only occasionally stirs himself on a Wednesday – means that all the transactions take place on a Wednesday, so when the balance becomes zero it is always on a Wednesday. So far, so good. But why do the zeros occur only at noon?

Titchmarsh wants to be sure that he's got it right, so he hires an assistant, an itinerant Polish accountant by the name of Odlyzko, to make a thorough investigation of the paperwork. Day after day, Odlyszko checks the zeros back into the mists of time. There seems to be no regularity about their distribution, apart from the 'critical time', as it's called. You can never know in which week you will find a zero; you can prove why, when it does occur, it must be on a Wednesday; but you've no idea why it is always at noon. At the back of Odlyzko's mind is the possibility that if he looks for long enough he will find a zero somewhere else. He knows it will never be on another day of the week, but if it's at, say, 10.13 a.m. or 6.22 p.m., it will totally confound Titchmarsh, and Odlyzko would be quite happy to do that, since he thinks Titchmarsh isn't paying him enough to compensate for the long hours and the eye strain.

That's an attempt to describe the state of play with the Riemann Hypothesis through the twentieth century, since mathematicians first seriously started to calculate zeros. Bernhard Riemann himself (compare our Mr Riemann, the bank customer) proved that the first few zeros of his function (the Zeta Bank account) fell within a narrow band of values (only on Wednesdays). He suspected that they lay only on a specific line, called the critical line, within that band (at noon). But he was unable to prove it. The same goes for everybody else ever since, even though every effort made to calculate yet more zeros, principally by Andrew Odlyzko, shows them all to lie on the critical line.

So the objective is clear – proving the Riemann Hypothesis is proving that a series of values of a particular function lie on a particular line. Some mathematicians work at this by finding another line and shifting it to coincide with the critical line; others by finding a random matrix with associated eigenvalues which exactly reproduce the Riemann zeros; others work at finding a function which, if you apply it over and over again to a particular curve or set of curves, produces different numbers of points that don't change. Over and over again, mathematicians from an increasingly wide range of mathematical fields have devised methods which have at their heart the Riemann zeta function but approach it by very different routes.

'Sometimes I think that we essentially have a complete proof of the Riemann Hypothesis except for a gap,' Hugh Montgomery said to me. 'The problem is, the gap occurs right at the beginning, and so it's hard to fill that gap because you don't see what's on the other side of it.'

Peter Sarnak said he was an optimist, but he sounded to me like a pessimist. 'One of the big differences here is that we're not competing with each other: we're really competing with what mankind can do. We don't know that this problem is anything that mankind can do right now – meaning, in my personal belief, that there's still a number of things that have to fall into place. Some things *are* beginning to fall into place. Certainly I'm an optimist. I hope this problem would be solved in the next ten or twenty years and then we might look back and say, "Now we understand why this couldn't have been done without this development and this development."'

Atle Selberg, the acknowledged doyen of 'Riemannology', said to me at the AIM Palo Alto workshop, 'At my age, I rather doubt that I will see a solution. That doesn't say that much. The problem is now nearly one and a half centuries old, and it is not impossible that it will become two centuries old before it is solved. It is possible that none of the people here at this conference would live to see a solution.'

According to the Oxford mathematician Roger Heath-Brown, 'Perhaps the most interesting thing is the resurgence of interest in the Riemann hypothesis over the past couple of years – amongst laymen, amongst specialists and amongst mathematicians from other areas. It's no longer just analytic number theorists involved, but all mathematicians know about the problem, and many realize that they may have useful insights to offer. As far as I can see, a solution is as likely to come from a probabilist,* geometer or mathematical physicist, as from a number theorist.'

For Yoichi Motohashi the Riemann Hypothesis is probably true, but he's prepared – as some are not – for the possibility that it is false. 'I shall not be surprised if the Riemann Hypothesis is proved – or disproved, though this is much more unlikely – in a couple of years. I mean by this that the Riemann Hypothesis will be settled without any fundamental changes in our mathematical thoughts, namely, all tools are ready to attack it but just a penetrating idea is missing.'

Henryk Iwaniec doesn't believe that anyone today has a very strong or convincing programme. 'Never mind a complete proof but a programme, a direction to go in. There are lots of analogous results that are similar, and people try to search through analogies, and that's interesting, but I don't know why it should be through analogy... I'm not convinced.'

But Iwaniec was certain of one thing: 'Mother Nature has such beautiful harmonies, so you couldn't say that something like that is false.'

But it could be. A few mathematicians are prepared for that possibility, and one or two even think it is likely. Of course, its falsity is easier to prove than its truth. One single Riemann zero that is off the critical line would completely destroy the hopes of those who are trying to prove it true.

There is a third theoretical possibility. In mathematics there is such a thing as a statement that is undecidable in any formal system. If this principle applies to the Riemann Hypothesis, it would mean that the

---

* A mathematician who specializes in probability theory.

hypothesis can never be proved to be either definitely true or definitely false. That statement – that the Riemann Hypothesis can never be proved definitely true or definitely false – would itself be provably true, according to something called Gödel's theorem, though no one believes that this is the case, least of all Alexander Ivic.

'I know for sure that the Riemann hypothesis is not undecidable,' he said, 'otherwise the whole of classical analysis would collapse. And classical analysis makes the world go round – the thing by which bridges are built up, planes fly, electricity flows, plastic is made – and so it's inconceivable that there may be a mathematics in which the Riemann Hypothesis would be false and another mathematics in which it will be true, like the axiom of parallels and non-Euclidian geometry. I cannot imagine universes, one in which this thing will be false and another in which it would be true. I'm not a philosopher – I'm a working mathematician, and I hope to stand firmly on the ground.'

Ivic mused on a question he is sometimes asked: whether he believes in the Riemann Hypothesis. 'I'm not in the business of religion,' he said. 'You may believe in God, for example. The Hindu religion believes in reincarnation. Now, I'm Serb Orthodox and my religion does not believe in reincarnation. No Christian Church believes in reincarnation. However, in discussing this with Hindus I'm very careful, because how the hell do I know? Maybe if I say I don't believe in reincarnation I will come back as a rabbit in the next life, God knows. I cannot say with certainty who is wrong. Someone must be wrong. Whether it's us or them or both…maybe it's undecidable, the question of reincarnation. So I don't believe or disbelieve the Riemann hypothesis. I have a certain amount of data and a certain amount of facts. These facts tell me definitely that the thing has not been settled. Until it's been settled it's a hypothesis, that's all. I would like the Riemann hypothesis to be true, like any decent mathematician, because it's a thing of beauty, a thing of elegance, a thing that would simplify many proofs and so forth, but that's all.'

But while the twenty or thirty mathematicians who are capable of proving the Riemann Hypothesis flounder in a miasma of frustration and uncertainty, one mathematician, Louis de Branges, has the confidence that escapes the rest of them. Since November 2000, he has acquired some more students – six of them, not merely the faithful Yasho – in a course he teaches on functional analysis.

'It's just wonderful to talk to people,' he said. They say, "I don't understand this," or it's clear that *I* don't understand it, and then we talk about it, and I finally understand what I'm supposed to be understanding!'

At the same time, he has continued to work away at his proof. When I spoke to him in the autumn of 2001 he had recently sent a manuscript to Nikolai Nikolski in Bordeaux. But this wasn't his absolutely final, ts crossed and i s dotted proof, it was more an indication of his approach to enable Nikolski to provide an independent validation of a National Science Foundation grant that de Branges needed for his next trip to France.

In the spring of 2002 de Branges received the result of his grant application. The panel of assessors, all of them anonymous, gave a range of ratings, from very good through good to fair, and there were two who gave his application no rating at all. But the final verdict was a firm 'No'. The NSF said that his proposal was 'non-competitive in the current funding environment'. The comments of the mathematicians who assessed de Branges's application ran the gamut from grudgingly charitable to mildly vitriolic.

One assessor wrote: 'The PI [Principal Investigator] is perhaps best known for his celebrated solution of the Bieberbach conjecture, and his description of prior work refers to (as yet unpublished) solutions of the Riemann hypothesis and the invariant subspace problem. I am not familiar enough with this work to comment on its correctness, but if true this would be a most remarkable accomplishment. The proposal itself is a brief description of issues related to the Riemann Hypothesis and the need to view the proof of the Bieberbach Conjecture in this context... It seems impossible to evaluate the PI's claims regarding the Riemann hypothesis... At present, support may not be justified.'

Another assessor said, 'De Branges has claimed to have solved the R.H. and the invariant subspace problem on several occasions in the past, but the proofs turned out to be lacking. There is no evidence that the current claims are more valid. His most recent paper cited in *Math Reviews* was published in 1994. I recommend against funding his proposal.'

A third, who rated the proposal as 'good', nevertheless wrote that 'The author has for over twelve years been producing versions of a paper claiming to prove this... The proposal suggests that there is a 2001 version of his Riemann series. I haven't looked at that one. The proposer's work on the Bieberbach Conjecture is absolutely brilliant. It is only on that basis that one continues try accommodate [sic] his continued claims of solutions of major unsolved problems. One must admire his courage to tackle these problems (after all that is a necessary condition for solving them). On the other hand given his track record of mistakes, his expectation that people should spend their time verifying his work is unreasonable.'

'This is probably the most controversial proposal I have ever reviewed,' said another assessor. 'Louis de Branges became famous for his proof of the Bieberbach Conjecture, which makes it unpleasant to deny him NSF support. On the other hand the present proposal provides insufficient evidence to guarantee success or even substantial progress.'

My interpretation of these remarks is that de Branges made it easier than it might otherwise have been for the panel to reject his ideas by supplying a very sketchy account of his work, though he knew that more detailed accounts were already available to the panel on the Internet. With the submission of much more detailed proposals from other, younger, mathematicians, with all the mathematical workings laid out for the panel to read, it was almost inevitable that they should reject de Branges's unpolished offering as 'non-competitive'.

Even with no NSF funding, de Branges went to France in May 2002 to present two seminars on his ideas at the Institut Henri Poincaré in Paris. He wrote me a letter, handwritten as usual, to tell me about the first meeting on 13 May.

'My lecture was carefully prepared on three large boards which moved electrically. I spoke in French... Tatiana told me afterwards that my voice is not as loud in French as it is in English, but that it was perfectly audible in view of the small audience.' (There were nine mathematicians present.)

As usual, de Branges expected his audience to do some work. 'The verification [of the proof] should cause no difficulty for anyone who is familiar with the theory of Hilbert spaces of entire functions,' he wrote. 'The difficulty is however that the members have no acquaintance with the Hilbert spaces, much less with these particular spaces. The seminar can only confirm the proof of the Riemann Hypothesis if it is first willing to learn the essentials... This might be done in several weeks if the seminar is willing to make the needed effort.'

De Branges gave a second seminar on 27 May, and I was there to hear it. I found him at 10:15 a.m. in the Institut's lecture room, a spacious auditorium two storeys high, with twenty or so rows of seats on a steeply raked floor. He was writing out the essence of his ideas in French on five blackboards. He hadn't quite finished by the time the audience assembled. This time there were ten mathematicians, seven men and three women. The faithful Tatiana sat discreetly at the back. The chairman of the session, a French Tunisian, Professor Daboussi, stood up in the row where he was sitting and introduced de Branges with a perfunctory sentence or two. De Branges then talked fluently for forty minutes, with a couple of hesitations as he realized he had written a plus instead of a minus or made some other trivial mistake. Two people, a man and a woman, wrote copious notes while he talked. The rest just sat, looking neither bored nor fascinated. By 11.40, de Branges had reached the end of the fifth board, and started to add some more material on the sixth. He was setting out the steps of his proof and indicating where there were still gaps.

'There is a complication that has caused me much difficulty,' he said. 'I have had a lot of trouble finding the right form. Maybe after five or six more seminars,' he said hopefully, addressing the group, 'we can find it.' There was no obvious show of assent. 'For Bieberbach we needed five,' he went on, and added encouragingly, 'and I think this is far simpler.'*

After the talk, a bearded, ascetic-looking mathematician asked several questions which suggested that he wasn't entirely familiar with de Branges's ideas but was interested in following them up. The rest of the audience slipped away with no comment.

I followed Daboussi and asked him whether it was likely that de Branges would get his 'five or six more seminars'. He shook his head. 'He says everything two or three times,' he said irritably. 'It's all too complicated, not easy to understand.'

Over lunch afterwards, I asked de Branges whether he felt there was anything else he could do to persuade the world that his 'proof' – when, or if, it was completed – would be worth looking at. The question seemed unimportant to him.

---

* The essence of de Branges's approach to a proof of the Riemann Hypothesis as laid out in the Paris lectures of May 2002 is given in the Appendix, for those mathematicians familiar with Hilbert spaces of entire functions.

'Why should anyone to try to do that?' he asked. 'The proof will be on the Internet. Anyone who is interested can read it. When somebody does decide to validate it I will get the credit.'

'But we all have only a few more decades to live,' I said. 'Don't you want it to be validated before you die?'

'*I* do,' said Tatiana.

After lunch, I asked him what he was going to do now. 'I'm going to walk a bit,' he said, and with unexpected consideration for his wife's views, he added, 'as much as Tatiana will let me.' Then he put on the beret that had so amused Michael Berry and set off with Tatiana through intermittent showers, for a stroll along the streets of the Left Bank.

Whatever his shortcomings, de Branges shares one characteristic with the mathematicians who disparage or ignore him. Like them, he is gripped by something which in some sense doesn't exist – at least, not in the way that he himself exists, or this book you hold in your hands exists. It is an idea, first conceived in the mind of a diffident German mathematician a century and a half ago, that won't go away. Like most – perhaps all – of the other mathematicians working on the problem, the reason de Branges spends most of his waking hours on this idea is not for money, and only a little for fame, but mainly because he believes it is true and desperately wants to prove it. In fact, he and the others might even die happy if they knew that *someone* had proved it.

'I would be happy to dig ditches, earn a million, and give it to somebody else for the proof,' Charles Ryavec told me. 'If I solved it, they could have the million. I would probably buy books for the students.'

Meanwhile, the antennae are waving in the breeze as the world's mathematicians e-mail one another or mingle in bars at conferences. 'We need a big idea,' they say, 'and no one has got it.' But however much they believe that, they dream of a day that most of them know will come, perhaps even in their lifetimes.

'The minute somebody is even sniffing, everybody knows,' said Peter Sarnak. 'It'll spread like wildfire, there's no question.'

Alain Connes told me a fable that encapsulates the devilish nature of the Riemann Hypothesis and its capacity to surprise and provoke. 'There's this mathematician who fought with this problem for many years, perhaps twenty years. At the end, of course, he is very frustrated but he is extremely curious about the answer, because after all it could fail – we don't know – it could be, for instance, that 100 per cent of the zeros are on the critical line, but there are a few exceptions... So he decides to do whatever is in his power to know the answer, and for this he is ready to sell his soul to the Devil. So he gets an appointment with the Devil. Of course, the Devil is a clever man, so he comes in with the papers to be signed first. So the guy signs the papers, and once he is done he asks the Devil, "Is the Riemann Hypothesis true?" The Devil looks at him and says, "What is the Riemann Hypothesis?" So then the guy begins to explain: "You take the inverse of the integers and raise them to power z, sum it when the real part of z is bigger than 1, and analytically continue to the critical strip, and then you want to know where the zeros are." And the Devil says, "I didn't know about this problem – it will take me some time to think about it. Let's meet in three days at the same place at midnight." Fine, though the guy is really taken aback – he has signed the papers and he doesn't yet know the answer. So he goes away, waits eagerly for three days and then comes back to the same place at midnight. There's nobody there. He waits half an hour. Still nobody. One o'clock in the morning – nobody shows up. Half past one. Nobody shows up. So eventually at a quarter past two, the Devil arrives sweating, all dishevelled, and he comes and he says, 'I couldn't do it but I could prove a nice little lemma*!'

Connes laughed gleefully. 'The moral of the story is that it is somehow a devilish problem. There are several possibilities, assuming, as we all believe, that it is true. It is probably at such a level, such a depth, that it would be a tragedy if it just needed a trick to prove it, something equivalent to taking a metal bar a hundred metres high, say, and balancing it vertically on its end. Something quite unbelievable but possible. And that such a proof would say nothing about this understanding of the adelic line. That would be tragic. On the other hand, it could also be that there is no way out other than really understanding the structure in full. And this would be great, but it means that we are still at an unknown distance from the target and we can only be patient. It would be quite tragic if there was a trick but it could happen, one never knows. We would feel very let down.'

---

* A subsidiary result proved in order to simplify the proof of a theorem.

TOOLKITS

TOOLKIT 1: **Logarithms and exponents**

When I was at school, before the days of electronic calculators, tables of logarithms were used to carry out complicated calculations. The way it worked was that if you had two big numbers you wanted to multiply together, you would first look up the logarithm of each number in a book of tables, usually sprinkled liberally with the graffiti of previous generations of schoolboys. This would give you two other numbers, and the next step was to add those numbers together. Finally, you looked up that new number – the result of the addition – in another part of the book of tables, under the heading 'antilogarithms', and the corresponding antilogarithm would be the answer you were looking for.

You could describe it like this. To multiply A by B, find L_A and L_B (the logarithms of A and B) and add them together. $L_A + L_B$ gives you a new number, L_C. This number is the logarithm of a number C. And C, which you find from looking up the table of antilogarithms, is actually $A \times B$. Its usefulness was clear – you could turn multiplication, a difficult process, into addition, an easier one.

To many people who were taught this rigmarole it seemed like magic, but as long as it worked it didn't matter, I suppose. But logarithms are not merely a useful tool. They crop up in many fields of mathematics, and they tell us something about the way things grow or shrink. When Gauss wanted to find a description of that gentle decline in the numbers of primes, he thought logarithms might come into it somewhere.

So what is the connection between a number and its logarithm, between A and L_A? And how can *adding* logarithms help you *multiply* numbers?

Here's how it works. The logarithms we used at school were based on the number 10. Our number system is called the decimal system because the way we write large numbers is based on powers of ten. A number as familiar as 100, for example, is written this way because it actually represents $(1 \times 10^2) + (0 \times 10) + (0 \times 1)$, whereas 111, a hundred and eleven, represents $(1 \times 10^2) + (1 \times 10) + (1 \times 1)$. The rule is that each digit has to be multiplied by a power of ten that increases as we read from the right.

In a large number such as 76,385,247,122, the 7, ten digits from the right, is actually multiplied by 10^{10}, and added to 6 times 10^9, which is added to 3 times 10^8, and so on.

In this system, the logarithm of A is the power that 10 has to be raised to in order to produce A. (The power of a number is also called its exponent.) So, to take a very simple example, the logarithm of 1,000 is 3. This is because if you raise 10 to the power 3 – multiply it by itself three times – you get 1,000. And the logarithm of 10,000 is 4. So if you wanted to find $A \times B$ using logs, L_A would be 3, L_B would be 4, and L_C – the logarithm of $A \times B$ – would be $L_A + L_B = 3 + 4 = 7$, which is the logarithm of the result of multiplying 1,000 by 10,000. And what number has 7 as its logarithm? The number 10 to the power 7: 10^7, i.e. 10,000,000. With this simple example we've found that $1,000 \times 10,000 = 10,000,000$, which we could have done by straightforward multiplication.

But those tables of logarithms had logarithms for all numbers, not just whole-number powers of ten. You could look up, say, 3.162 277 66 and find that the log was 0.5. How can this be? Literally, it means that if you raise 10 to the power 0.5 the result is 3.162 277 66, but it's not easy to see what raising something to a non-integer power means. How can you multiply 10 by itself 0.5 times?

One of the most fruitful ways of advancing mathematics in the past has been by extension – by taking an idea that works within a certain framework and asking what would happen if we extended the framework to include a wider range of possibilities. Whoever first wondered what it could mean if there were such things as exponents that are not whole numbers, 10 to the power half, for example, was using just that kind of extension of a basic framework and opened up a whole new field in the process.

The use of exponents, raised numbers such as ^{2}, ^{3} and ^{10} that indicate how many times a number is to be multiplied together, follows fairly simple rules. When you multiply together two numbers such as x^a and x^b, the answer is x^{a+b}. It's easy to see why: x^7 times x^4 is x^{11} because $(x \times x \times x \times x \times x \times x \times x) \times (x \times x \times x \times x)$, when you remove the brackets, is just 11 xs multiplied together. We can see that in action in $10^3 \times 10^4 = 10^7$.

When people started looking closely at exponents, they asked whether there was any meaningful way of extending their use beyond positive whole numbers. One extension that seemed as though it might be useful was to ask what it would mean to raise a number to the power 0. It turned out that the exponent 0 could be incorporated into the system. Just as, say, $7^3 \times 7^5 = 7^{3+5} = 7^8$, we could try using the same rule to multiply, say, 7^3 by 7^0 – even if we don't yet know what 7^0 means – and see what we get. The answer is $7^3 \times 7^0 = 7^{3+0} = 7^3$. What this shows is that 7^0 is the number that gives you 7^3 when you multiply 7^3 by it. That is, the number that leaves another number unchanged when you multiply by it. In other words, if we take 7^0 as equal to 1 we can then extend our system of numbers to incorporate the power 0.

Of course, there's nothing special about 7 in this example. The same reasoning will show that any number to the power 0 works exactly as the number 1 does in any equation. So if $x^0 = 1$, what sort of meaning can we arrive at for x^1? We can try the same trick, with the same example. Let's see what we get if we multiply 7^3 by 7^1. Well, $7^3 \times 7^1 = 7^{3+1} = 7^4$. So a workable meaning for 7^1 is the number that produces 7^4 when we multiply 7^3 by it. In this case, it's clear that to get 7^4 from 7^3 we have to multiply 7^3 by 7. So a useful meaning for 7^1 is 7 – and, to generalize, $x^1 = x$.

Now the big step. So far we've just worked our way back from obvious whole-number exponents such as squares and cubes to less obvious ones, 1 and 0. But supposing we try to find a meaning for exponents that are not whole numbers, but fractions or decimals. What does something like $7^{1/2}$ mean?

If we try exactly the same process as before, we get $7^3 \times 7^{1/2} = 7^{7/2}$, which doesn't get us much further since we don't know what on earth $7^{7/2}$ means. But we can try a variant of that trick that gives us an answer we can understand. Suppose that we multiply $7^{1/2}$ by $7^{1/2}$ using the addition of powers. Well, $7^{1/2} \times 7^{1/2}$ must equal 7^1, i.e. $7^{1/2} \times 7^{1/2} = 7$. By this reasoning, we can take $7^{1/2}$ to mean the number which, when multiplied by itself, equals 7. And this reasoning says that $7^{1/2}$ therefore equals the square root of 7.

That's what fractional exponents are taken to mean, and they lead to an extension of the system that works very well. $7^{1/3}$ is the cube root of 7. $7^{1/10}$ is the tenth root of 7, and so on. You can even have something like $7^{2/3}$. So trying $7^{2/3} \times 7^{1/3} = 7$ suggests that $7^{2/3}$ is the number that when multiplied by the cube root of 7 gives you 7. In other words, $7^{2/3}$ is the cube root of 7, squared.

We started by asking how the logarithm of a number could be a decimal – how 3.162 277 66 could be ten raised to the power 0.5. Well, since 0.5 is $\frac{1}{2}$, the answer is in our grasp. 10 raised to the power 0.5 is the same as $10^{1/2}$ and this, as with the example above, means the square root of ten. So the log of 3.162 277 66 is 0.5, which means that 3.162 277 66 is the square root of ten, which we can easily check. To multiply 3.162 277 66 by 3.162 277 66 we take logs and add them. Adding 0.5 to 0.5 gives 1, and so 3.162 277 66 × 3.162 277 66 is the number whose log is 1, i.e. 10. I picked an easy exponent, $\frac{1}{2}$, but in fact we can raise 10 – or any other number – to any decimal power, so that every positive number can be expressed as a power of 10. And that power for each number is its logarithm.

In these examples we've used logarithms based on powers of 10 (we say that 10 is the base of the logarithms), but it's possible to have other systems with any base we choose. Using logs to the base 2, for example, would mean that the log of 4 is 2, the log of 8 is 3, the log of 16 is 4, and so on, because these logs are the power to which 2 has to be raised to produce the particular number.

Mathematicians sometimes talk about a particular process increasing logarithmically. Out at sea, for example, the speed of wind over water increases logarithmically with height. That means that if the wind speed is *v* at 10 metres, it is *v* times log 2 at twice that height, 20 metres, and *v* times log 4 at 40 metres. The logarithms of numbers increase much more slowly than the numbers themselves do – while the numbers go from 1 to 10 to 100, the logs go from 1 to 2 to 3, so wind speed rises much more slowly with increasing height.

TOOLKIT 2: **Equations**

Equations are like evenly balanced scales. The pivot is the equals sign, and you start with two sets of quantities or mathematical expressions which are in the pans on either side of the pivot and which balance out. In the simplest type of equation, the pans contain one or more unknown weights all of the same value and a lot of known weights, and your job is to find the value of the unknown weight. So if the unknown weight, x, is in the left pan with eleven other weights each of 1 kilo, and it is balanced by 15 kilos in the right pan, you can guess that the unknown weight is 4 kilos. But guesses can only get you so far in maths. Mathematicians like methods and proofs. The method in this case is to find how to get the x alone in one pan and balanced by some number of kilos in the other pan.

You can probably see that if you carry out some operation on one pan, adding or subtracting weights, for example, the scales will remain balanced only if you do the same operation on the other pan. Adding a 100 kilos to one pan will severely unbalance the scales; adding 100 to both will have no effect. So, to return to our simple example, if we want to have just the unknown weight in the left pan, we have to remove 11 kilos. To keep the scales balanced we also have to remove 11 kilos from the right hand pan. This leaves 4 kilos of the original 15. So we now have x on one side balanced by 4 kilos on the other, so x must equal 4.

This is a laborious but perhaps helpful way of representing the solution of the equation $x + 11 = 15$. The mathematical way of dealing with this equation is by subtracting 11 from both sides to get $x = 4$. When you find a value for x, that value is called the root of the equation – the number that makes both sides equal.

If we had, say, three of the unknown weights in one pan along with 7 kilos, and they were balanced by 16 kilos in the other pan, it's slightly more complicated. We can start by taking 7 kilos from each pan, which leaves 9 kilos on one side but with three unknown weights on the other. But on the principle that we can do what we like to one side as long as we do it to the other, we can then divide the contents of each pan by 3 and it should still balance. This leaves us with x in one pan and 3 kilos in the other, and we've discovered that the unknown weight is 3 kilos. So, in symbols,

$$3x + 7 = 16$$
$$3x = 9$$
$$x = 3$$

The key procedure is always to carry out identical operations on both sides of the equation so as to isolate the unknown, x, in such a way that its value is on the other side of the equation.

There's a simpler way of doing this. It looks different, but it gives exactly the same results. With $x + 11 = 15$, for example, instead of subtracting 11 from both sides, you can achieve the same effect by moving 11 from the left side of the equation to the right, and changing its sign from $+$ to $-$:

$$x + 11 = 15$$
$$x + \textcircled{11} = 15 \,\overline{\nwarrow}$$
$$x = 15 - 11$$
$$x = 4$$

This process is summed up in a phrase that you may remember from school maths lessons: 'change side, change sign'. You can move a number from one side of the equation to the other, and the two sides will balance as long as you change the sign, from plus to minus or minus to plus. In a similar way, instead of, say, dividing both sides of $3x = 9$ by 3, you can 'move' the 3 from where it is multiplying x, and use it to divide the other side, changing the sign from 'times' to 'divides', thus:

$$3x = 9$$
$$\textcircled{3}x = \underline{9} \,\nearrow$$
$$x = \frac{9}{3}$$
$$x = 3$$

These manipulations, and many others like them, are second nature to people who work with equations all the time. For any equation containing an unknown in its simplest form, x, these methods are enough to solve them. But equations can be written in x^2, x^3 or higher powers of x, and these require very different methods to solve them – if indeed it is possible to solve them at all. Equations in x^2 are called quadratic equations, and those in x^3 are called cubic equations.

With all these equations, the aim is to get the unknown on one side and everything else on the other, but there's not usually a standard method of doing so.

Another point that's important is that equations in x^2 or higher powers have more than one root. An equation with x^2 in it has two roots, a cubic has three roots, and so on. As an example, here's a quadratic equation:

$$x^2 + x = 12$$

It has two roots, so there are two values of x which make it work: x can equal 3 or –4. If it's 3, we get $3^2 + 3 = 12$, which is true, and if we try –4, we get $(-4)^2 - 4 = 12$, which is also true.

This equation can also be expressed as $(x-3)(x+4) = 0$. You'll see why if we multiply the brackets out, using a fairly simple procedure. If you have two sets of brackets with terms in them, you take the first term in the first bracket and multiply it by each of the terms in the second bracket, one by one. Then you take the second term in the first bracket and do the same thing. Finally you add all the terms together. So, in the above example, where x is first term in the first bracket and –3 is the second, we have

$$x \times (x + 4) \; = \; x^2 + 4x$$

and

$$-3 \times (x + 4) \; = \; -3x - 12$$

Multiplying out and adding all the terms together gives

$$x^2 + 4x - 3x - 12 \; = \; x^2 + x - 12$$

So we've shown that $(x-3)(x+4)$ is another way of expressing $x^2 + x - 12$, and they both equal 0. To say that $x^2 + x - 12$ equals 0 is the same as saying that $x^2 + x = 12$ (by moving the 12 to the other side of the equation and changing its sign from + to −), which is the equation we started with.

There's one other point to notice here. If the equation had first been written as $(x-3)(x+4)=0$, we would have seen very easily what its roots are. You should be able to see straight away that putting x equal to 3 will make the first bracket equal 0 and the equation will be satisfied (because zero times anything equals zero). Similarly, putting x equal to −4 in the second bracket will have the same effect.

Infinity has to be handled carefully in mathematics. Jack Littlewood, a brilliant English mathematician who appears often in this book, described the following paradox:

> Balls numbered 1,2, … (or for a mathematician the numbers themselves) are put into a box as follows. At 1 minute to noon the numbers 1 to 10 are put in, and the number 1 is taken out. At $\frac{1}{2}$ minute to noon numbers 11 to 20 are put in and the number 2 is taken out. At $\frac{1}{3}$ minute to noon [the numbers] 21 to 30 in and 3 out; and so on. How many are in the box at noon? The answer is none: any selected number, e.g. 106, is absent, having been taken out at the 106th operation.[120]

Part of the problem lies in treating infinity as an actual number, a slipshod practice which is useful only when we are discussing mathematics loosely. But this practice is full of pitfalls. For example, the whole numbers form an infinite series, and so do the prime numbers – they both go on for ever, yet they can't both have the same 'number' of members (called terms) since one series is included in the other. But it's still possible to play with infinite series in various ways and get meaningful results. Mathematicians like adding or multiplying the terms together – something that you might think will often produce infinity as a result. But early mathematicians soon realized that this wasn't always so. Certainly, $1+2+3+4+5+6+7+\ldots$ to infinity adds up to an infinitely large number, as does $1+3+5+7+11+13+17+19+\ldots$ (the sum of the series made up of all the prime numbers). But how about $1+\frac{1}{2}+\frac{1}{4}+\frac{1}{8}+\frac{1}{16}+\ldots$? It turns out that however many terms you add to this series, the total never exceeds 2. And this not just because the terms are fractions. Some series made up of adding together an infinite number of fractions *do* add up to an infinitely large number. $\frac{1}{2}+\frac{1}{3}+\frac{1}{4}+\frac{1}{5}+\frac{1}{6}+\frac{1}{7}+\frac{1}{8}+\frac{1}{9}+\ldots$ is one of them. (This is just one of those surprises that maths springs on you all the time.) In fact, you need quite sophisticated mathematical tools to discover which series *diverge* (the totals get larger and larger as you add more terms) and which *converge* (no matter how many terms you add, the total never exceeds a certain number).

This infinite series of fractions is very important to our story:

$$\frac{1}{1^2} + \frac{1}{2^2} + \frac{1}{3^2} + \frac{1}{4^2} + \frac{1}{5^2} + \frac{1}{6^2} + \frac{1}{7^2} + \cdots$$

It's a series of reciprocals. In mathematics, a reciprocal of a number is simply 1 divided by that number. So the reciprocal of 2 is a half, the reciprocal of 32 is $\frac{1}{32}$, and so on. You can make similar series that involve higher powers than just *squaring* the numbers on the bottom. You can cube them:

$$\frac{1}{1^3} + \frac{1}{2^3} + \frac{1}{3^3} + \frac{1}{4^3} + \frac{1}{5^3} + \frac{1}{6^3} + \frac{1}{7^3} + \cdots$$

or you can use the 19th power of each integer:

$$\frac{1}{1^{19}} + \frac{1}{2^{19}} + \frac{1}{3^{19}} + \frac{1}{4^{19}} + \frac{1}{5^{19}} + \frac{1}{6^{19}} + \frac{1}{7^{19}} + \cdots$$

Whatever you choose as the power of the numbers on the bottom, the result is still an infinite series.

An individual term in the series is described in general as $1/n^s$, where n takes the value of each whole number in turn, and s is the power to which each number is raised. This may already be one mathematical symbol too many for some readers, but it's important to get to know a few very specific symbols because they will keep on cropping up.

Symbols whose meaning is not as well known as $+$ and $-$ are simple to understand when they are explained, because, even though they may not familiar, the concepts they describe are. Σ, for example, a large Greek capital sigma, means 'sum', and is applied to lists of elements that have to be added together.

Let's look at another way of writing one of the series of fractions we met earlier. Instead of putting $1 + \frac{1}{2} + \frac{1}{4} + \frac{1}{8} + \frac{1}{16} \ldots$, we could write:

$$\sum \frac{1}{2^n} \quad \text{for} \quad n = 0, 1, 2, 3, 4, \ldots$$

This expression means 'the sum of the terms where the bottom of each fraction is 2 raised to a power that goes up by one with each term'. This is because the bottom parts of the fractions are 2, 2^2, 2^3, 2^4, ... So you could work out what any term of the series is by substituting the appropriate n, which would make, say, the third term $1/2^3$, i.e. $\frac{1}{8}$, and the tenth term $1/2^{10}$. (2^{10}, remember, is $2 \times 2 \times 2 \times 2 \times 2 \times 2 \times 2 \times 2 \times 2 \times 2$, i.e. 2 multiplied together ten times.)

Here's another series that looks vaguely similar:

$$\sum \frac{1}{n^s}$$

Here, as before, the n indicates the natural numbers, increasing by one in each term. So you could write this out as:

$$\frac{1}{1^s} + \frac{1}{2^s} + \frac{1}{3^s} + \frac{1}{4^s} + \frac{1}{5^s} + \cdots$$

But what's the s that's crept in here? Well, it just shows that this particular sum of fractions has increasing natural numbers on the bottom of each fraction, with each number raised to the same power. So if s is 2, we get the sum

$$\frac{1}{1^2} + \frac{1}{2^2} + \frac{1}{3^2} + \frac{1}{4^2} + \frac{1}{5^2} + \cdots$$

In other words,

$$\frac{1}{1} + \frac{1}{4} + \frac{1}{9} + \frac{1}{16} + \frac{1}{25} + \cdots$$

So, remembering our original notation, we have a series of fractions added together (but going on for ever) which could be described much more succinctly as $\sum 1/n^s$, where $n = 1, 2, 3, 4, 5, \ldots$ and $s = 2$.

This doesn't seem too much to get excited about, but in fact the expression $\sum 1/n^s$ is at the heart of the Riemann Hypothesis. This group of symbols has a specific value obtained by adding all the terms of the series together, and that value varies as different values of s are substituted. So although it looks like the description of a single series, it's really shorthand for a whole family of series, each one based on a different value of s. (Remember, the n just means that each series has all the natural numbers up to infinity on the bottoms of the fractions.)

TOOLKIT 4: **The Euler identity**

One of the most extraordinary features of the zeta function – the original Euler zeta function that doesn't use complex numbers – is a relationship that Leonhard Euler himself discovered. You'll remember that the zeta function is based on adding the reciprocals of all the whole numbers raised to a certain power. So to find the value of $\zeta(5)$, zeta for the power 5, for example, you write out $1^5, 2^5, 3^5, 4^5, 5^5, \ldots$, then write the sequence again as reciprocals to get $1/1^5, 1/2^5, 1/3^5\ 1/4^5, 1/5^5, \ldots$ for ever. If you now add these reciprocals together you get a single number (which is not infinitely large because these numbers get smaller and smaller in a certain way that leads to a finite sum) which is the value of $\zeta(5)$.

$\zeta(s)$ for any whole number s is expressed mathematically as

$$\sum_{n=1}^{\infty} \frac{1}{n^s}$$

and you say this as 'the sum of one over n to the s for n from one to infinity'. What Euler discovered was that, for any s, the value of $\zeta(s)$ can also be given by an entirely different expression that uses only prime numbers. There are several important differences between this new expression and the sum involving all the integers. The first is that where the classic zeta function involves terms of the form $1/n^s$, with n taking successive values of the integers, Euler's new expression uses $1/p^s$, with p taking only successive prime number values. The second difference is that the $1/p^s$ is included in a slightly more complicated term:

$$\frac{1}{1 - \dfrac{1}{p^s}}$$

And the third difference is that each term, as p goes through all the primes, is *multiplied* by the other terms rather than added to them.

Euler's new expression is written as

$$\prod_{p=primes} \frac{1}{1 - \dfrac{1}{p^s}}$$

where the symbol Π, a large capital Greek pi (not to be confused with π, the lower-case pi that stands for the constant 3.141 59…), means 'the product of'. Euler said that this product is equal to

$$\sum_{n=1}^{\infty} \frac{1}{n^s}$$

for all values of s.

To make it easier to understand by dropping the Σs and Πs, here is what Euler said:

$$\frac{1}{1^s} + \frac{1}{2^s} + \frac{1}{3^s} + \frac{1}{4^s} + \frac{1}{5^s} + \cdots$$

$$= \left(\frac{1}{1 - \dfrac{1}{2^s}} \right) \left(\frac{1}{1 - \dfrac{1}{3^s}} \right) \left(\frac{1}{1 - \dfrac{1}{5^s}} \right) \left(\frac{1}{1 - \dfrac{1}{7^s}} \right) \left(\frac{1}{1 - \dfrac{1}{11^s}} \right) \cdots$$

This is called an *identity*, because it is true for all values of s – unlike an equation, which is true only for certain values of the unknown.

Euler's discovery has been called 'one of the most remarkable discoveries in mathematics'. It is remarkable because it is so surprising. On one side of the equals sign is a series of terms increasing in a stately fashion, with each n increasing by 1 on the bottom of the fractions. And we understand the ns very well – they are the familiar whole numbers we use to count with. On the other side of the equals sign is an expression where each successive term contains a number drawn from the list of prime numbers – which, in Euler's time even more than today, were much more poorly understood than the integers. Although we can't even write a function that will generate each successive prime number, they seem to be intimately related to the integers by Euler's discovery.

Why is this so? If you're with me so far, you might think it worth sticking around to find out. If we take one term of the product on the right, we have

$$\frac{1}{1-\dfrac{1}{p^s}}$$

We can simplify this to

$$\frac{p^s}{p^s-1}$$

by expressing the bottom fraction as

$$\frac{p^s-1}{p^s}$$

and then turning it upside down, since the reciprocal of a fraction is just the fraction turned upside down. (So 1 divided by $\frac{1}{2}$ is just $\frac{1}{2}$ turned upside down, which is 2.)

To understand the next step, it's easier to work backwards from the answer. Let's multiply the expression p^s-1 (you'll see why in a minute) by the following series:

$$1 + \frac{1}{p^s} + \frac{1}{p^{2s}} + \frac{1}{p^{3s}} + \frac{1}{p^{4s}} + \cdots$$

To multiply these two expressions together, we just have to multiply the whole series first by p^s and then by -1 and add the results together. Multiplying by p^s produces the following:

$$p^s\left(1 + \frac{1}{p^s} + \frac{1}{p^{2s}} + \frac{1}{p^{3s}} + \frac{1}{p^{4s}} + \cdots\right)$$

$$= p^s + 1 + \frac{1}{p^s} + \frac{1}{p^{2s}} + \frac{1}{p^{3s}} + \frac{1}{p^{4s}} + \cdots$$

And multiplying the same series by -1 gives us:

$$-1 - \frac{1}{p^s} - \frac{1}{p^{2s}} - \frac{1}{p^{3s}} - \frac{1}{p^{4s}} - \cdots$$

If we now add together the results of multiplying first by p^s and then by -1, the second multiplication cancels out most of the result of the first one and leaves p^s.

So, to recap. We showed that if we multiply a series, which we'll call A, by $p^s - 1$, we got the answer p^s. So $A(p^s - 1) = p^s$. This means that

$$\frac{p^s}{p^s - 1} = A$$

which is the series

$$1 + \frac{1}{p^s} + \frac{1}{p^{2s}} + \frac{1}{p^{3s}} + \frac{1}{p^{4s}} + \cdots$$

But as we saw above,

$$\frac{p^s}{p^s - 1}$$

is one term of the Euler identity, so we can rewrite the whole identity as

$$\sum_{n=1}^{\infty} \frac{1}{n^s} = \left(1 + \frac{1}{2^s} + \frac{1}{2^{2s}} + \frac{1}{2^{3s}} + \frac{1}{2^{4s}} + \cdots\right) \times$$

$$\left(1 + \frac{1}{3^s} + \frac{1}{3^{2s}} + \frac{1}{3^{3s}} + \frac{1}{3^{4s}} + \cdots\right) \times$$

$$\left(1 + \frac{1}{5^s} + \frac{1}{5^{2s}} + \frac{1}{5^{3s}} + \frac{1}{5^{4s}} + \cdots\right) \cdots$$

and so on through all the primes.

We are now getting a little nearer to proving that the right side equals the left. The next step seems (to me at any rate) formidable. We now have to multiply together this infinite number of brackets, each of which contains an infinite number of terms. And all of this before next Christmas.

But what Euler knew was the following fact: every whole number is the result of multiplying together a unique group of prime numbers (this is the Fundamental Theorem of Arithmetic). So 2,324,168 $= 2^3 \times 7^4 \times 11^2$, for example, and that number cannot be obtained by multiplying together any other combination of primes. What he spotted was that, out of the mammoth multiplication of every term in each bracket by every term in every other bracket, you would get a series of numbers added together, each of which was of the form 1 divided by a series of prime numbers raised to certain powers, thus:

$$\frac{1}{p_1^{as}\ p_2^{bs}\ p_3^{cs}\ p_4^{ds}\ p_5^{es}\ p_6^{fs}\ p_7^{gs}\ p_8^{hs}\ p_9^{is}\ p_{10}^{js}\ p_{11}^{ks}\ p_{12}^{ls}\cdots}$$

Here, $p_1, p_2, p_3, \ldots$ are the primes in succession, and they will be multiplied together with every possible combination of exponents, because each term in each bracket will multiply every other term in every other bracket.

Finally, Euler saw that each of these terms on the bottom of the series of fractions will equal one of the natural numbers n, and that they would all be different, since every n was equal to a unique combination of prime numbers multiplied together. This, the product of terms involving all the primes, has been turned into the sum of a series of terms involving all the whole numbers.

One of the major advances in mathematics was the discovery that it is useful to represent algebraic expressions as lines on a graph.

The simplest graphs have a horizontal axis and a vertical axis at right angles to it. Numbers are marked off at equal intervals along these axes, with positive numbers going to the right and up, and negative numbers to the left and down. The horizontal axis is called the x-axis, and the vertical one the y-axis. In this system, any point can be represented by two numbers, x and y. If x is 2 and y is 3, this refers to a point which is 3 units above the horizontal line and 2 to the right of the vertical line. Now, if we were to mark on this graph only those points where the x and the y were related mathematically in some way, say points for which $3x = y$, we would find that those points all form a straight line, starting at the zero point where the axes intersect, called the origin, and sloping up to the right. If the relationship between x and y was different, say $2x = 7y$, we'd get a different line, still passing through the origin but having a different slope. How do we know that this line passes through the origin? Well, ask yourself what is the value of y when $x = 0$, and you'll see that $y = 0$ too. But what if we take an expression like $4x + 3$ and set it equal to y? This turns out to make a straight line too, but now, when $x = 0$, $y = 3$, so this straight line still slopes up to the right, but doesn't pass through the origin. Instead it cuts the y-axis where $y = 3$.

You can plot any mathematical function with two variables as a line or curve on a flat graph. The expression $x^2 + y^2 = 1$, which is the same as $x^2 = 1 - y^2$, describes a circle, because all the points that satisfy this relationship are to be found at a distance of one unit from the point of origin of the graph.

Where all this is leading to is that there is a graphical way of representing the Euler zeta function by plotting a series of points on a graph with s along the horizontal axis and $\zeta(s)$ along the vertical. Then, if $s = 2$, say, $\zeta(s) = \Sigma \, 1/n^2$. Or, to expand it,

$$\zeta(2) = 1 + \frac{1}{2^2} + \frac{1}{3^2} + \frac{1}{4^2} + \frac{1}{5^2} + \cdots$$

the first five terms of which add up to 1.463 611 111..., and when you add all the other terms to infinity you get 1.644 934 067.*

This process allows us to put points on a graph where the horizontal coordinates are values of s and the vertical coordinates are values of $\zeta(s)$. $\zeta(3)$ is $1 + 1/2^3 + 1/3^3 + 1/4^3 + 1/5^3 + ...$, and these terms add up to 1.185 662 037.... .

In a similar way, we can fill in the gaps between whole-number values of s with fractional values, so that, for example,

$$\zeta\left(1\tfrac{1}{2}\right) = 1 + \frac{1}{2^{1\frac{1}{2}}} + \frac{1}{3^{1\frac{1}{2}}} + \frac{1}{4^{1\frac{1}{2}}} + \frac{1}{5^{1\frac{1}{2}}} + \ ...$$

Since $2^{1\frac{1}{2}}$ is $\sqrt{(2^3)}$, i.e. $\sqrt{8}$, the series starts $1 + 1/2.828\,427\,125$, and you can probably see how to work out the rest of it to arrive at a value for $\zeta(1\tfrac{1}{2})$.

An interesting thing happens when s is 1, because we get the infinite series $1 + \tfrac{1}{2} + \tfrac{1}{3} + \tfrac{1}{4} + \tfrac{1}{5} + ...$, which actually grows larger and larger without limit as you add more and more fractions, so $\zeta(1)$ is infinite.

So for real values of s, i.e. 'ordinary numbers', we can work out the values of $\zeta(s)$ and plot them like a line on a graph, as in Figure 19. What

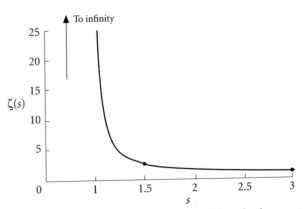

FIGURE 19 When s is a real number, the function $\zeta(s)$ can be drawn as a two-dimensional graph, by taking values of s along the x axis and working out a value for $\zeta(s)$ on the y axis. Here, very schematically, the points on the line mark $\zeta(3)$ and $\zeta(1.5)$. But because $\zeta(1)$ produces a value on the y axis that is infinitely large, the graph shoots up to infinity when $s = 1$.

---

* This turns out in a mysterious way to equal exactly $\pi^2/6$, but we'll let that pass.

we have been able to do with the *two* variables is to represent how the function behaves as a graph on a *two*-dimensional sheet of paper, by creating two axes at right angles, which we could think of as north–south and east–west. One of the variables is taken as marking the distance from the horizontal axis and the other from the vertical, uniquely specifying a value of $\zeta(s)$ for any real number s. But the *Riemann* zeta function has four variables. This is because the values of s we are now interested in are complex numbers. That is, each s has to be written in the form $x+iy$. And when we work out $\zeta(x+iy)$ we get another complex number, say $u+iv$. It is not possible to show the relationships between these four variables on a two-dimensional graph. It's not even possible in three dimensions. In fact, to do it properly we need four-dimensional graph paper.

If we want to try to show the Riemann zeta function in graphical form, we have to work out a way of plotting on a graph those xs, ys, us and vs so that a single point is represented by each unique combination of the four variables. To extrapolate from the two-dimensional situation, we'd need four axes each at right angles to the other three, something that is not possible in our own familiar three dimensions (known to mathematicians as Euclidean space), positions in which we can represent graphically with reference to three axes at right angles to one another: east–west, north–south and up–down. In this space we can represent every point uniquely with three separate coordinates, but we've got *four* coordinates: x, y, u and v. So to represent the behaviour of the Riemann zeta function for all complex values we need a four-dimensional graph, something difficult to visualize in our obstinately three-dimensional world, though it seems to present no problems for mathematicians.

The best we can do in trying to understand the Riemann zeta function in a graphical way is to find a means of representing the value of $\zeta(x+iy)$, which is a number like $u+iv$, as one figure rather than two, and as a figure whose value can go to zero when $\zeta(x+iy)$ equals zero. Figure 20 shows how this is possible. If z is the complex number $x+iy$, we can give it a single value by calculating the expression $\sqrt{(x^2+y^2)}$ and, since x and y are real numbers, we get just one real number to go with u and v on the other two axes.

To recap, we have devised a way in which, to any value of $\zeta(s)$ (a function involving a complex number), we can assign a single real number,

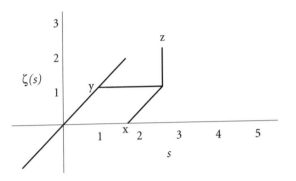

FIGURE 20 If we wanted to plot the zeta function $\zeta(s)$ where $s = x + iy$, we get an answer of the form $u + iv$. But to plot $\zeta(x + iy) = u + iv$ as a graph we'd need to do it in four dimensions because it involves four variables, x, y, u and v.

However, mathematicians have devised a way of 'collapsing' the graph into three dimensions by representing the complex number $u + iv$ as a single real number z, represented by the number $\sqrt{(u^2 + v^2)}$. So if we work out $\zeta(x + iy)$ for any values of x and y on the horizontal axes we will get a complex answer, $u + iv$, and using that answer we can work out a value for z and put it as a point z units above the x,y plane.

which means that we can work out a value for $\zeta(s)$ for all values of u and v, and this is a third number, z. With three variables we can plot all values of $\zeta(s)$ on a three-dimensional graph. Clearly we've lost something by condensing the x and the y and representing them as one number z, because we've produced a certain ambiguity. Instead of a unique value for every $\zeta(s)$, some different values of x and y will produce the same z. Since z is $\sqrt{(x^2 + y^2)}$, x could be +2 or −2, and y could be +3 or −3, and any combination of those would produce the same values of z. But we can at least construct a 'picture' of the Riemann zeta function that is not complete but holds useful information, rather like a three-dimensional 'shadow' of a four-dimensional object. That's how mathematicians have arrived at the representation of the Riemann zeta function shown in Figure 20.

Matrices are mathematical tools for representing assemblies of numbers which need to be treated as a single mathematical entity. They can have any size, but a 2×2 is one of the simplest. It looks like this:

$$\begin{pmatrix} a & b \\ c & d \end{pmatrix}$$

where a, b, c and d, called the elements of the matrix, stand for numbers, real or complex. Matrices don't have to be square. They can be rectangular – a 6×15 matrix, for example, with six rows and fifteen columns – or a single column or row of numbers. Matrices that have a practical use in physics may have thousands or even millions of rows and columns, and it's easy to see why a way of handling them as single objects is very useful. There are ways to add and subtract, multiply and divide these objects, and these methods usually boil down to conventional operations with the individual elements. But the results of matrix multiplication, say, are far from conventional for those of us used to certain consequences in ordinary multiplication. 2×3 always gives the same result as 3×2 when we are dealing with everyday numbers, but with matrices that is not necessarily the case.

These operations are done in a special way. Multiplication of two 2×2 matrices, for example, is done as follows:

$$\begin{pmatrix} a & b \\ c & d \end{pmatrix} \times \begin{pmatrix} e & f \\ g & h \end{pmatrix} = \begin{pmatrix} ae + bg & af + bh \\ ce + dg & cf + dh \end{pmatrix}$$

If we summarize this as $A \times B = C$, we'll see that applying the same rules to $B \times A$ doesn't give the same result:

$$\begin{pmatrix} e & f \\ g & h \end{pmatrix} \times \begin{pmatrix} a & b \\ c & d \end{pmatrix} = \begin{pmatrix} ea + fc & eb + fd \\ ga + hc & gb + hd \end{pmatrix}$$

So changing the order of multiplication from $A \times B$ to $B \times A$ produces not C, but a different matrix (D, say). This may seem strange when compared with conventional multiplication, but in fact other ordinary arithmetical operations aren't always 'commutative' (as it is called when the order doesn't matter). Take division, for example: 11 divided by 7 is not the same as 7 divided by 11.

Matrices have specific numbers in each position, usually chosen for some specific purpose such as solving equations. For example, you might take the following three equations, and try to find the values of x, y and z that would solve them together:

$$2x + 5y + 5z = 1$$
$$4x + 10y = 2$$
$$x + 10y = 8$$

you could represent those equations as three matrices like this:

$$\begin{pmatrix} 2 & 5 & 5 \\ 4 & 10 & 0 \\ 1 & 10 & 0 \end{pmatrix} \times \begin{pmatrix} x \\ y \\ z \end{pmatrix} = \begin{pmatrix} 1 \\ 2 \\ 8 \end{pmatrix}$$

There are ways of manipulating this matrix to find the values of x, y and z. (Like any algebraic manipulation of an equation, it's a matter of transforming both sides of the equation by subjecting them to the same procedure.) It's done by turning the 3×3 matrix into an equivalent matrix with 1s along the diagonal (don't worry about how) so that the expression above is transformed into the equivalent expression

$$\begin{pmatrix} 1 & 0 & 0 \\ 0 & 1 & 0 \\ 0 & 0 & 1 \end{pmatrix} \times \begin{pmatrix} x \\ y \\ z \end{pmatrix} = \begin{pmatrix} -2 \\ 1 \\ 0 \end{pmatrix}$$

If you imagine turning this back into three equations, you'll see that it's equivalent to $x = -2$, $y = 1$ and $z = 0$.

This may seem complicated, but these examples give a hint of the

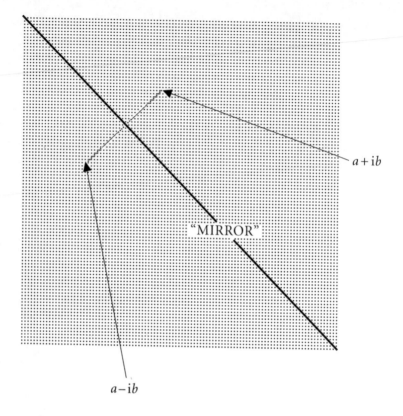

$a + ib$

"MIRROR"

$a - ib$

FIGURE 21 One type of random matrix, called a self-adjoint matrix, holds a clue to the Riemann Hypothesis. The diagonal of this matrix is a kind of mirror, reflecting one half of it in the other in a mathematical way. you can choose any complex number, say a + ib, and place it anywhere in the half of the matrix to the right of the diagonal. The rule for self-adjoint matrices says that you then have to place the 'conjugate' of that number, obtained by changing the sign in the middle from plus to minus or minus to plus, in a corresponding position on the other side of the diagonal.

complexity of the sort of matrices that are conventionally used in mathematics and physics. In aviation, for example, an engineer might want to understand the fluttering of an aeroplane's wing. To do this, she would typically use matrices that are 1,000 rows by 1,000 columns. Just to multiply two of them together would mean that for each element there are a thousand additions, and there are a million elements altogether.

The matrices that Michael Berry and Jon Keating work with, random matrices, differ in three fundamental ways from the simple examples above. First, their elements include complex numbers; second, these numbers are selected randomly but in a rather particular way; and third, the matrices have an infinite number of rows and columns.

Such strange objects have emerged in mathematics because they reflect the underlying nature of some problems in physics for which there are very large numbers of variables that need to be manipulated in certain ways. One type of random matrix, called a self-adjoint matrix, holds a clue to the Riemann Hypothesis. The diagonal of this matrix is a kind of mirror, reflecting one half of it into the other in a mathematical way. 'Reflecting' in this context means that each element on one side of the diagonal has a partner in a corresponding position on the other side, chosen according to some particular rule. With real numbers, you could imagine picking a number at random for the bottom left corner of the matrix and putting its reciprocal, for example, in the top right corner. In random matrices, the elements are all complex numbers. So a complex number is chosen at random, $a + ib$, say, and a number called its complex conjugate is put an equal distance from the diagonal but on the other side ('reflected'), as shown in Figure 21.

To find the 'complex conjugate' of a number $a + ib$ (remember these are complex numbers), you just change the plus to a minus, so the conjugate of $a + ib$ is $a - ib$. A number and its conjugate have some interesting properties. One of them is that if you multiply them together you get a real number with no imaginary part. Remembering that i is treated as the square root of minus one, we get $(a + ib)(a - ib) = a^2 + iab - iab - i^2b$. The iabs cancel out and, since $i^2 = -1$, the whole thing boils down to $a^2 + b^2$, which is a real number.

So far, then, we have an infinite matrix with one side of the diagonal filled with randomly chosen numbers, and the other side filled with conjugates in the corresponding positions. Then, to complete the matrix, the diagonal is filled with real numbers, again randomly chosen. This is how you build a self-adjoint random matrix, and Figure 22 shows a very diagrammatic representation of an infinite self-adjoint random matrix.

There's one other characteristic of matrices that's important in this story. Each $n \times n$ matrix has a set of n special numbers associated with it.

These numbers are called eigenvalues. The 2×2 matrix

$$\begin{pmatrix} 1 & 2 \\ 1 & 3 \end{pmatrix}$$

has two eigenvalues. They happen to be the roots of an equation that can easily be derived from the matrix, in this case a quadratic equation because the matrix is 2×2. The equation for this matrix is $x^2 - 4x + 1 = 0$, and so the eigenvalues are $2 - \sqrt{3}$ and $2 + \sqrt{3}$. These work out to be approximately 0.267 949 192 and 3.732 050 808. A $1,000 \times 1,000$ matrix would have a thousand eigenvalues, the thousand roots of an equation starting with $x^{1,000}$.

You can think of the eigenvalues as a unique 'spectrum' of the matrix.

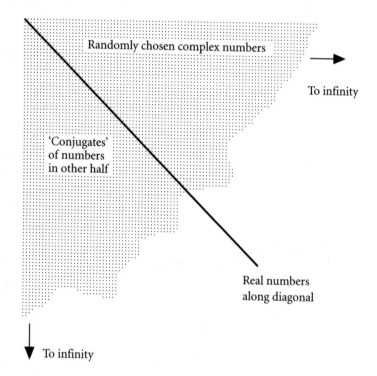

FIGURE 22 The random matrices which are important for the Riemann Hypothesis are self-adjoint (see Figure 23) and have an infinite number of rows and columns. Mathematicians deal with a set of these matrices which has an infinite number of members.

Chemical elements and molecules have spectra, consisting of a series of distinct lines of different wavelengths. They can be written down as a series of numbers, and each element has a unique spectrum by which it is identified.

But random-matrix theory is not content with just one matrix, with about half its elements chosen at random. The theory deals with a whole collection of random matrices. Having chosen the numbers arbitrarily to fill half of this infinite matrix, we could have made a number of different choices to fill those spaces. In fact, we could make an infinite number of different matrices, by choosing different numbers.

'You have infinite matrices,' said Michael Berry, 'and you have infinitely many of them, a vast infinity of them. Now this thing is called the random-matrix ensemble, this collection, and it has a rather high order of infinity.'

There's one more step to take before we can see the connection between random matrices and the Riemann zeta function. Each of the infinite matrices in the infinitely large ensemble has a list of eigenvalues associated with it, and there are an infinite number of eigenvalues in each list, each list being called the spectrum of the matrix.

When random matrices are used to analyse the behaviour of systems in quantum physics, this spectrum of eigenvalues corresponds to energy levels, because quantum mechanics is based on the fact that in the microscopic world – at the level of very small entities – physical properties such as energy can only exist in 'packets', called quanta, unlike the smoothly changing aspects of the macroscopic world, governed as it is by classical mechanics.

Berry summed it all up: 'The values that the energy levels of an electron in an atom can take are the eigenvalues of a matrix. And the way you compute the energy levels of atoms and molecules is to find the right matrix and find its eigenvalues.'

These mysterious quantities known as eigenvalues arise in the following way (see Figure 23). Suppose we're interested in the eigenvalues of a square matrix A. If we take a different matrix B – not square this time, but one that consists of a single column – and multiply it by A, the result will be another single-column matrix C, different from B. But there will be some special single-column matrices B – special in their relationship with A – for which multiplication by A generates a C that reproduces B exactly, except for being scaled up or down by some numerical

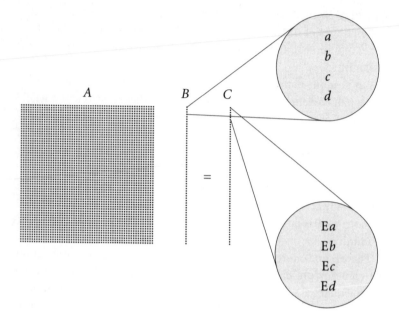

FIGURE 23 With any $n \times n$ matrix A, it's possible to find an $n \times 1$ matrix B such that when A and B are multiplied together they produce a matrix C that is a 'scaled' version of B, i.e. all the elements of B multiplied by the same number. Matrices that behave like B are called eigenvectors of A, and the 'scaling factor' (E) is called an eigenvalue of A.

factor. And there as many single-column matrices, B_1, B_2, B_3, B_4, ..., which have this special property as there are rows or columns in the matrix. The matrices (B_1, B_2, B_3, B_4, ...,) are called eigenvectors of A, and the numerical 'scaling' factors associated with each eigenvector are called eigenvalues.

To summarize: a matrix – an array of numbers arranged in rows and columns – has associated with it a series of numbers called eigenvalues, n of them if it's an $n \times n$ matrix. With a matrix that has an infinite number of rows and columns there will be an infinite series of numbers which are the eigenvalues. For the infinite number of matrices in the infinite ensemble of random matrices there are an infinite number of sets of infinite eigenvalues. Now here's the punchline – for one of those infinite matrices, the set of eigenvalues may correspond exactly to the zeros of the Riemann zeta function.

NOTES

All unattributed quotations in the text are from interviews conducted by the author.

1. Louis de Branges de Bourcia, 'Apology for proof of the Riemann Hypothesis', http://www.math.purdue.edu/~branges/

2. Constance Reid, *Julia, A Life in Mathematics*, Mathematical Association of America, Washington DC, 1996, p. 54

3. Clay Mathematics Institute, 'The Riemann Hypothesis', http://www.claymath.org/prizeproblems/riemann.htm

4. Constance Reid, *Julia, A Life in Mathematics*, Mathematical Association of America, Washington DC, 1996, p. 19

5. Paolo Ribenboim, *The New Book of Prime Number Records*, Springer-Verlag, New York, 1996, pp. 159–63

6. G.H. Hardy, 'The Sixth Josiah Willard Gibbs Lecture', read at New York City, 28 December 1928 before a joint session of the American Mathematical Society and the American Association for the Advancement of Science. Reprinted in *Bulletin of the American Mathematical Society*, 1929, **35**, pp. 778–818

7. Reinhard Laubenbacher and David Pengelley (eds), *Mathematical Explorations: Chronicles by the Explorers*, Springer-Verlag, New York, 1999, p. 187

8. André Weil, *The Apprenticeship of a Mathematician*, translated by Jennifer Gage, Birkhäuser, Basel, 1992, p. 145

9. James Parton, *Sir Isaac Newton*. Quoted in Robert Édouard Moritz (ed.), *Memorabilia Mathematica*, Macmillan, New York, 1914, Quote 1022

10. A. N. Whitehead, *Introduction to Mathematics*, New York, 1911, pp. 59–60. Quoted in Robert Édouard Moritz (ed.), *Memorabilia Mathematica*, Macmillan, New York, 1914, Quote 1218

11. Sir Arthur Eddington, quoted in N. Rose (ed.), *Mathematical Maxims and Minims*, Raleigh, NC, Rome Press, 1988

12. Bertrand Russell, quoted in N. Rose (ed.), *Mathematical Maxims and Minims*, Raleigh, NC, Rome Press, 1988

13. Johann Goethe, quoted in J. R. Newman (ed.) *The World of Mathematics*, Simon & Schuster, New York, 1956, p. 1754

14. Maria Price La Touche, Letter to the *Mathematical Gazette*, 1878, Vol. 12. Quoted in Donald E. Knuth, *The Art of Computer Programming*, Vol. 2, Addison-Wesley, Reading, MA, 1997

15. Leonhard Euler, quoted in Howard Eves, *Mathematical Circles*, Prindle, Weber, & Schmidt, Boston, 1969

16. David Brewster, *Letters of Euler*, Vol. 1, New York, 1872, p. 24. Quoted in Robert Édouard Moritz (ed.), *Memorabilia Mathematica*, Macmillan, New York, 1914, Quote 959

17. Leonhard Euler, quoted in G. Simmons, *Calculus Gems*, McGraw-Hill, New York, 1992

18. Ruth McNeill, 'A Reflection on When I Loved Maths and How I Stopped.' *Journal of Mathematical Behavior*, 1988, **7**, 45–50

19. I.M. Gelfand and A. Shen, *Algebra*, Birkhäuser, Boston, 1995

20. Reinhard Laubenbacher and David Pengelly (eds), *Mathematical Expeditions: Chronicles by the Explorers*, Springer-Verlag, New York, 1999, p. 136

21. Robert Musil, *Young Törless*, Secker & Warburg, 1955

22. Marjorie A. Tiefert, The Samuel Coleridge Taylor Archive, maintained by the University of Virginia Library at http://etext.lib.virginia.edu/stc/Coleridge/ascii_files/geometry_poem_letter.html

23. *Ibid.*

24. *Life of Benjamin Robert Haydon, Historical Painter, From His Autobiography and Journals*, 3 vols, edited and compiled by Tom Taylor (London, 1853)

25. *Ibid.*

26. John Keats, *Lamia*, pt. ii, lines 229–37

27. James Boswell, *Life of Johnson*, Harper's Edition, 1871. Quoted in Robert Édouard Moritz (ed.), *Memorabilia Mathematica*, Macmillan, New York, 1914, Quote 981

28. J.S.C. Abbott, *Napoleon Bonaparte*, Vol. 1, New York, 1904, Chap. 10. Quoted in Robert Édouard Moritz (ed.), *Memorabilia Mathematica*, Macmillan, New York, 1914, Quote 1001

29. John Aubrey, 'Thomas Hobbes', in *Aubrey's Brief Lives* (ed. Oliver Lawson Dick), Oxford University Press, 1960, p. 604

30. E. Lampe, *Die Entwickelung der Mathematik* ..., Berlin, 1893, p. 22. Quoted in Robert Édouard Moritz (ed.), *Memorabilia Mathematica*, Macmillan, New York, 1914, Quote 1115

31. Artie Shaw, *The Trouble with Cinderella: An Outline of Identity*, Fithian Press, 1992, p. 273

32. G.H. Hardy, *Ramanujan*, Cambridge University Press, 1940, p. 5

33. E.T. Bell, *The Development of Mathematics,* Dover Publications, New York, 1992, p. 369

34. Constance Reid, *Hilbert*, Springer-Verlag, New York, 1996, p. 73

35. David Hilbert, *Bulletin of the American Mathematical Society*, 1902, **8**, 437–45, 478-479. Quoted in Reid, *Ibid.*, p. 75

36. *Ibid.*, p. 83

37. Constance Reid, *Julia: A Life in Mathematics*, Mathematical Association of America, Washington, DC, 1996, pp. 69–73

38. Bela Bollobas, foreword to *Littlewood's Miscellany*, Cambridge University Press, 1986, p. 16

39. G.H. Hardy, *Ramanujan*, Cambridge University Press, 1940, p. 22

40. C.P. Snow, foreword to G.H. Hardy, *A Mathematician's Apology*, Cambridge University Press, 1967, p. 48

41. Harald Bohr, *Collected Mathematical Works in Three Volumes*, edited by Erling Følner and Børge Jessen, Dansk Matematisk Forening, Copenhagen, 1953, pp. xxvii–xxviii

42. J.E. Littlewood, *Littlewood's Miscellany*, edited by Bela Bollobas, Cambridge University Press, 1986, p. 76

43. *Ibid.*, p. 87

44. *Ibid.*, p. 85

45. *Ibid.*, p. 88

46. Bela Bollobas, *Ibid.*, foreword, p. 16

47. G.H. Hardy, quoted in Robert Kanigel, *The Man Who Knew Infinity*, Scribners, New York, 1991

48. J.E. Littlewood, *Littlewood's Miscellany*, edited by Bela Bollobas, Cambridge University Press, 1986, p. 61

49. G.H. Hardy, *A Mathematician's Apology*, Cambridge University Press, 1967

50. Rae Chorze Fwaz Mystery School, 'Adventures in Self Discovery', http://www.rae-chorze-fwaz.com

51. David Ash, 'Rama, Ramanujan, and 1729', http://www.rae-chorze-fwaz.com/1729.html

52. Charles Simonyi, EDGE 29, 19 November 1997, http://www.edge.org/documents/archive/edge29.html

53. Stanislas Dehaene, EDGE 29, 19 November 1997, http://www.edge.org/documents/archive/edge29.html

54. J.E. Littlewood, *Littlewood's Miscellany*, edited by Bela Bollobas, Cambridge University Press, 1986, pp. 97–9

55. G.H. Hardy, *Mathematical Gazette*, 1910, **9** (5), 333–4

56. Gerald Alexanderson, *The Random Walks of George Pólya*, Mathematical Association of America, Washington, DC, 2000, p. 72

57. 'Math appreciation', *CCS Notes*, June 1994, pp. 10–11

58. The description of this extraordinary day is taken from Sylvia Nasar, *A Beautiful Mind*, Faber & Faber, London, 1998, p. 245–6

59. *Ibid.*, p. 246

60. Harold M. Edwards, *Riemann's Zeta Function*, Dover Publications, New York, 2001, p. ix

61. Alan Hodges, *Alan Turing: The Enigma*, Vintage, London, 1992, p. 140

62. *Ibid.*, p. 156

63. Jacques Hadamard, 'Sur la Distribution des zéros de la fonction $\zeta(s)$ et ses conséquences arithmétiques', *Bulletin de la Société Mathématiques de France*, 1896, **14**, 199–220

64. Joe Shipman, 'De Branges claims to have shown measurable cardinals inconsistent', http://www.math.psu.edu/simpson/fom/postings/9907/msg00024.html (19 July 1999)

65. Rev. J. Spence, *Anecdotes, Observations, and Characters of Books and Men*, London, 1858, p. 132. Quoted in Robert Édouard Moritz (ed.), *Memorabilia Mathematica*, Macmillan, New York, 1914, Quote 1010

66. Paul Hoffman, *The Man Who Loved Only Numbers*, Fourth Estate, London, 1998, pp. 192–3

67. C. J. Keyser, *Lectures on Science, Philosophy, and Art*, New York, 1908, p. 29. Quoted in Robert Édouard Moritz (ed.), *Memorabilia Mathematica*, Macmillan, New York, 1914, Quote 515

68. E.J. Borowski and J.M. Borwein, *Dictionary of Mathematics*, HarperCollins, London, 1989

69. Alfred Adler, 'Reflections: Mathematics and creativity', *New Yorker*, 1972, **47** (53), 39–45

70. G.H. Hardy, *Nature*, 1934, **134**, 250

71. Albert Baernstein II *et al.* (eds), The Bieberbach Conjecture (Proceedings of the Symposium on the Occasion of the Proof), American Mathematical Society, Providence, RI, 1986, p. vii

72. *Ibid.*, p. 2

73. Walter Gautschi, in *Ibid.*, p. 205

74. *Ibid.*, p. 209

75. *Ibid.*, p. 213

76. *Ibid.*, p. 210

77. Louis de Branges, in *Ibid.*, p. 201

78. *Ibid.*, p. 201

79. *Ibid.*, p. 201

80. Richard Askey, in *Ibid.*, p. 214–15

81. Wim Klein, quoted in Stanislas Dehaene, EDGE 29, 19 November 1997, http://www.edge.org/documents/archive/edge29.html

82. Neil Sloane and Simon Plouffe, *The Encyclopedia of Integer Sequences*, Academic Press, San Diego, 1995. See also http://www.research.att.com/~njas/sequences/Seis.html

83. Malcolm Lines, *A Number for Your Thoughts*, Adam Hilger, Bristol, 1986, p. 30

84. G.H. Hardy, *Ramanujan*, Cambridge University Press, 1940, p. 17

85. Paul Hoffman, *The Man Who Loved Only Numbers*, Fourth Estate, London, 1998, pp. 40–41

86. A.M. Odlyzko, 'On the distribution of spacings between zeros of the zeta function', http://www.research.att.com/~amo/doc/arch/zeta.zero.spacing.pdf

87. Jean-Pierre Changeux and Alain Connes, *Conversations on Mind, Matter, and Mathematics*, edited and translated by M.B. DeBevoise, Princeton University Press, 1995, p. 50

88. A.S. Besicovitch, in J.E. Littlewood, *A Mathematician's Miscellany*, Methuen, London, 1953

89. Henri Poincaré, quoted in Jacques Hadamard, *The Mathematician's Mind*, Princeton University Press, 1996, p. 13

90. Gösta Mittag-Leffler, letter to Georg Cantor, 9 March 1885, quoted in Joseph Dauben, *Georg Cantor: His Mathematics and Philosophy of the Infinite*, Princeton University Press, 1990, p. 138

91. A.M. Legendre, 'Réflexions sur différentes manières de démontrer la théorie des parallèles ou le théorème sur la somme des trois angles du triangle', *Mémoires de l'Académie (Royale) des Sciences (de l'Institut de France)*, 2nd series, 1833, **12**, 367–410, plus sheet of figures

92. Reinhard Laubenbacher and David Pengelly (eds), *Mathematical Expeditions: Chronicles by the Explorers*, Springer-Verlag, New York, 1999, p. 26

93. J.E. Littlewood, *Littlewood's Miscellany*, edited by Bela Bollobas, Cambridge University Press, 1986, p. 93

94. Doron Zeilberger, 'Opinion 39: Partial and inconclusive proofs are welcome!' [= Elul 29, 5759], www.math.rutgers.edu/~zeilberg/Opinion39.html (10 September 1999)

95. Yuri Matijasevitch, in Constance Reid, *Julia, A Life in Mathematics*, Mathematical Association of America, Washington DC, 1996, p. 103

96. G.H. Hardy, The Sixth Josiah Willard Gibbs Lecture, read at New York City, 28 December 1928 before a joint session of the American Mathematical Society and the American Association for the Advancement of Science. Reprinted in *Bulletin of the American Mathematical Society*, 1929, **35**, 778–818

97. G.H. Hardy, *Ibid.*, pp. 778–9

98. Constance Reid, *Julia, A Life in Mathematics*, Mathematical Association of America, Washington, DC, 1996, p. 47

99. Charles Swartz, *Infinite Matrices and the Gliding Hump*, World Scientific, Singapore, 1996

100. Plutarch, *Marcellus*, translated by John Dryden (see http://classics.mit.edu/Plutarch/marcellu.html)

101. Bela Bollobas, foreword to J.E. Littlewood, *Littlewood's Miscellany*, Cambridge University Press, 1986, p. 17

102. Bertrand Russell, *The Study of Mathematics: Philosophical Essays*, London, 1910, p. 73. Quoted in Robert Édouard Moritz (ed.), *Memorabilia Mathematica*, Macmillan, New York, 1914, Quote 1104

103. J. J. Sylvester, 'Johns Hopkins Commemoration Day Address', in *Collected Mathematical Papers*, Vol. 3, Cambridge University Press, p. 76. Quoted in Robert Édouard Moritz (ed.), *Memorabilia Mathematica*, Macmillan, New York, 1914, Quote 1032

104. André Weil, *The Apprenticeship of a Mathematician*, translated by Jennifer Gage, Birkhäuser, Basel, p. 91

105. P. Turen, 'The Work of Alfréd Rényi', *Matematikai Lapok*, 1970, 21, 199–210

106. Paul R. Halmos, *I Want to be a Mathematician: An Automathography*, New York, 1985

107. Thomas Jefferson, letter to William Green Munford, 18 June 1799, quoted in J. Robert Oppenheimer, 'The encouragement of science', in I. Gordon and S. Sorkin (eds) *The Armchair Science Reader*, Simon & Schuster, New York, 1959

108. Jean-Pierre Changeux and Alain Connes, *Conversations on Mind, Matter, and Mathematics*, edited and translated by M. B. DeBevoise, Princeton University Press, 1995, pp. 75–6

109. C. C. Moore, 'The work of Alain Connes', *Notices of the American Mathematical Society*, 1982, **29**, 499–501

110. Press Release, Royal Swedish Academy of Sciences, 25 January 2001

111. J.E. Littlewood, *Littlewood's Miscellany*, edited by Bela Bollobas, Cambridge University Press, 1986, p. 59

112. André Weil, *The Apprenticeship of a Mathematician*, translated by Jennifer Gage, Birkhäuser, Basel, p. 182

113. Stephen Leacock, quoted in Howard Eves, *Return to Mathematical Circles*, Prindle, Weber, & Schmidt, Boston, 1988

114. Verena Huber-Dyson, 'On the nature of mathematical concepts: Why and how do mathematicians jump to conclusions?', http://www.edge.org/3rd_culture/huberdyson/huberdyson_p3.html

115. L.J. Mordell, *Three Lectures on Fermat's Last Theorem*, Chelsea Publishing, New York, 1962

116. Hermann Hankel, *Die Entwickelung der Mathematik in den letzten Jahrhunderten*, Tübingen, 1884, p. 9. Quoted in Robert Édouard Moritz (ed.), *Memorabilia Mathematica*, Macmillan, New York, 1914, Quote 718

117. Stobaeus, (Edition Wachsmuth, 1884), Ecl. I). Quoted in Robert Édouard Moritz (ed.), *Memorabilia Mathematica*, Macmillan, New York, 1914, Quote 952

118. Reinhard Laubenbacher and David Pengelly (eds), *Mathematical Expeditions: Chronicles by the Explorers*, Springer-Verlag, New York, 1999, p. 95

119. E. P. Wigner, 'The unreasonable effectiveness of mathematics in the natural sciences', *Communications on Pure and Applied Mathematics*, 13 February 1960

120. J.E. Littlewood, *Littlewood's Miscellany*, edited by Bela Bollobas, Cambridge University Press, 1986, p. 26

APPENDIX: **De Branges's proof**

The basis of Louis de Branges's proof of the Riemann Hypothesis, translated from the presentation at the Seminar in Number Theory at the Institut Henri Poincaré in Paris, May 2002.

The Riemann Hypothesis for Dirichlet Zeta functions

Louis de Branges de Bourcia

A Dirichlet zeta function

$$\zeta(s) = \sum \chi(n)n^{-s}$$

is defined in the half-plane $\mathbb{R}\, s > 1$ as a sum over the positive integers n with χ a primitive character modulo ρ other than the principal character modulo one. The zeta function has an analytic extension to the complex plane. The functional identity for the zeta function states that the entire functions

$$\left(\rho/\pi\right)^{\frac{1}{2}v+\frac{1}{2}s} \Gamma\!\left(\tfrac{1}{2}v + \tfrac{1}{2}s\right)\zeta(s)$$

and

$$\left(\rho/\pi\right)^{\frac{1}{2}v+\frac{1}{2}-\frac{1}{2}s} \Gamma\!\left(\tfrac{1}{2}v + \tfrac{1}{2} - \tfrac{1}{2}s\right)\zeta(1-s^-)^-$$

are linearly dependent with v equal to zero or one of the same parity as χ. The entire function

$$E(z) = \left(\rho/\pi\right)^{\frac{1}{2}v+\frac{1}{2}-\frac{1}{2}iz} \Gamma\!\left(\tfrac{1}{2}v + \tfrac{1}{2} - \tfrac{1}{2}iz\right)\zeta(1-iz)$$

satisfies the inequality

$$\left| E(x - iy) \right| < \left| E(x + iy) \right|$$

for y positive which permits the construction of an associated Hilbert space $\mathcal{H}(E)$ of entire functions. A maximal dissipative transformation in the space implies the simplicity of zeros of $E(z)$ and their position on the line $iz^- - iz = -1$. The space is constructed from Hilbert spaces of entire functions appearing in the theory of the Hankel transformation of order v.

The Laplace transformation of order v for the complex plane is defined when v is a nonnegative integer. The domain of the transformation is the set of functions $f(\xi)$ of a complex variable ξ which satisfy the identity

$$f(\omega\xi) = \omega^v f(\xi)$$

for every element ω of the unit circle and which are square integrable with respect to Lebesgue measure for the plane. The Laplace transform of order v of the function $f(\xi)$ of ξ in the plane is the analytic function $g(z)$ of z in the upper half-plane defined by the integral

$$2\pi g(z) = \int \left(\xi^v \right)^- f(\xi) \exp\left(\pi i \xi^- z \, \xi/\rho \right) d\xi$$

with respect to Lebesgue measure for the plane. When v is zero, the identity

$$\left(2\pi/\rho \right) \sup \int_{-\infty}^{+\infty} \left| g(x + iy) \right|^2 dx = \int \left| f(\xi) \right|^2 d\xi$$

holds with the least upper bound taken over all positive numbers y. The identity

$$\left(2\pi/\rho \right)^{1+v} \int_0^\infty \int_{-\infty}^{+\infty} \left| g(x + iy) \right|^2 y^{v-1} \, dx \, dy = \Gamma(v) \int \left| f(\xi) \right|^2 d\xi$$

holds when v is positive. Integration is with respect to Lebesgue measure for the plane.

A Hilbert space of functions analytic in the upper half-plane is obtained. The elements of the space are characterized by convergent integrals. A maximal dissipative transformation in the space is defined by taking $f(z)$ into $(i/z)^{\frac{1}{2}} f(z)$ when both functions of z belong to the space. The square root is taken with positive real part.

The domain of the Hankel transformation of order v for the complex plane is the domain of the Laplace transformation of order v for the complex plane. The Hankel transformation of order v for the complex plane is its own inverse and satisfies the identity

$$\int \left| f(\xi) \right|^2 d\xi = \int \left| g(\xi) \right|^2 d\xi$$

when it takes a function $f(\xi)$ of ξ in the complex plane into a function $g(\xi)$ of ξ in the complex plane. The identity

$$\int \left(\xi^v \right)^- g(\xi) \exp\left(\pi i \xi^- z \xi/\rho \right) d\xi$$
$$= \left(i/z \right)^{1+v} \int \left(\xi^v \right)^- f(\xi) \exp\left(-\pi i \xi^- z^{-1} \xi/\rho \right) d\xi$$

with z in the upper half-plane defines the Hankel transformation of order v for the complex plane. Integration is with respect to Lebesgue measure for the plane. A fundamental property of the Hankel transformation of order v for the complex plane was discovered in 1880 by Nikolai Sonine. If a is a positive number, a nontrivial function $f(\xi)$ of ξ in the complex plane exists which is in the domain of the Hankel transformation of order v for the complex plane, which vanishes in the neighborhood $|\xi| < a$ of the origin, and whose Hankel transform of order v for the complex plane vanishes in the same neighborhood.

The Mellin transformation of order v for the complex plane is a spectral theory for the Laplace transformation of order v for the complex plane. The domain of the transformation is the domain of the Laplace transformation of order v for the complex plane. If a function $g(z)$ of z in the upper half-plane is the Laplace transform of order v for the complex plane of a function $f(\xi)$ of ξ in the plane which vanishes when $|\xi| < a$, then its Mellin transform of order v for the complex plane is the analytic function $F(z)$ of z in the upper half-plane defined by the integral

$$F(z) = \int_0^\infty g(it) t^{\frac{1}{2}v - \frac{1}{2} - \frac{1}{2}iz} dt$$

Mellin transforms of order v for the complex plane are characterized using weighted Hardy spaces.

An analytic weight function is a function $W(z)$ of z in the upper half-plane which is analytic and without zeros in the half-plane. The weighted Hardy space $\mathcal{F}(W)$ is the set of analytic functions $F(z)$ of z in the upper half-plane such that a finite least upper bound

$$\sup \int_{-\infty}^{+\infty} \left| \frac{F(x+iy)}{W(x+iy)} \right|^2 dx$$

is obtained over all positive numbers y. Boundary values

$$\frac{F(x)}{W(x)} = \lim \frac{F(x+iy)}{W(x+iy)}$$

exist almost everywhere with respect to Lebesgue measure for the real line. The least upper bound is equal to the integral

$$\left\| F(t) \right\|^2_{\mathcal{F}(W)} = \int_{-\infty}^{+\infty} \left| \frac{F(t)}{W(t)} \right|^2 dt$$

with respect to Lebesgue measure for the real line.

The analytic weight function

$$W(z) = \left(\rho / \pi \right)^{\frac{1}{2}v + \frac{1}{2} - \frac{1}{2}iz} \Gamma\left(\frac{1}{2}v + \frac{1}{2} - \frac{1}{2}iz \right)$$

is applied in the characterization of Mellin transforms of order v for the complex plane. The Mellin transform of order v for the complex plane of a function $f(\xi)$ of ξ in the complex plane which vanishes when $|\xi| < a$ is an analytic function $F(z)$ of z in the upper half-plane such that the function

$$a^{iz} F(z)$$

belongs to the space $\mathcal{F}(W)$. Every element of the space $\mathcal{F}(W)$ is of this form. A maximal dissipative transformation in the space $\mathcal{F}(W)$ is defined by taking $F(z)$ into $F(z+i)$ whenever the functions $F(z)$ and $\mathcal{F}(z+i)$ of z belong to the space $\mathcal{F}(W)$. The set of entire functions

$F(z)$ such that $F(z)$ and

$$F^*(z) = F(z^-)^-$$

belong to the space $\mathcal{F}(W)$ is a Hilbert space which is contained iso-metrically in the space $\mathcal{F}(W)$. The elements of the space are the Mellin transforms of order v for the complex plane of functions in the domain of the Hankel transformation of order v for the complex plane which vanish in the neighborhood $|\xi| < a$ of the origin and whose Hankel transform of order v for the complex plane vanishes in the same neighborhood. The Sonine spaces of order v for the complex plane, which are so obtained, are fundamental examples of Hilbert spaces of entire functions which have an axiomatic characterization.

Consider a Hilbert space $\mathcal{H}$ whose elements are entire functions and which has these properties.

(H1) Whenever $F(z)$ belongs to the space and has a nonreal zero w, the function

$$F(z)\left(z - w^-\right)/\left(z - w\right)$$

belongs to the space and has the same norm as $F(z)$.

(H2) The linear functional on the space which takes $F(z)$ into $F(w)$ is continuous for every nonreal number w.

(H3) The function $F^*(z)$ belongs to the space whenever $F(z)$ belongs to the space and it always has the same norm as $F(z)$.

An example of such a space is obtained when $E(z)$ is an entire func-tion which satisfies the inequality

$$\left| E(x - iy) \right| < \left| E(x + iy) \right|$$

for y positive. A weighted Hardy space $\mathcal{F}(E)$ exists since $E(z)$ has no zeros in the upper half-plane. The space $\mathcal{H}(E)$ is the set of entire functions $F(z)$ such that $F(z)$ and $F^*(z)$ belong to the space $\mathcal{F}(E)$. The space $\mathcal{H}(E)$ is a Hilbert space which is contained isometrically in the space $\mathcal{F}(E)$. The space satisfies the axioms (H1), (H2), and (H3). The function

$$K(w,z) = \frac{E(z)\, E(w)^- - E^*(z)\, E(w^-)}{2\pi i(w^- - z)}$$

of z belongs to the space for every complex number w. The identity

$$F(w) = \left\langle F(t), K(w,t) \right\rangle_{\mathcal{H}(E)}$$

holds for every element $F(z)$ of the space. A Hilbert space whose elements are entire functions, which satisfies the axioms (H1), (H2), and (H3), and which contains a nonzero element, is isometrically equal to a space $\mathcal{H}(E)$.

The Sonine spaces of order v for the Euclidean plane are Hilbert spaces of entire functions which satisfy the axioms (H1), (H2), and (H3). The space of parameter a contains a nonzero element for every positive number a. A maximal dissipative transformation in the space is defined by taking $F(z)$ into $F(z+i)$ when the functions $F(z)$ and $F(z+i)$ of z belong to the space.

The Hilbert space of entire functions associated with a Dirichlet zeta function is also a Hilbert space of entire functions which satisfies the axioms (H1), (H2), and (H3). The space is a space $\mathcal{H}(E)$ with

$$E(z) = \left(\rho/\pi \right)^{\frac{1}{2}v + \frac{1}{2} - \frac{1}{2}iz} \Gamma\left(\frac{1}{2}v + \frac{1}{2} - \frac{1}{2}iz \right) \zeta(1 - iz)$$

The functional identity for the zeta function implies that the entire functions $E(z-i)$ and $E^*(z)$ are linearly dependent. A maximal dissipative transformation in the space exists which implies that the zeros of $E(z)$ are simple and lie on the line $iz^- - iz = -1$. A construction of the space $\mathcal{H}(E)$ from the Sonine spaces of order v for the complex plane is an interpretation of convergence of the Euler product. The maximal dissipative transformation in the space $\mathcal{H}(E)$ is derived from the maximal dissipative transformations in the Sonine spaces of order v for the complex plane.

The Euler product

$$\zeta(s)^{-1} = \prod \left(1 - \chi(p)p^{-s} \right)$$

for the zeta function is taken over the primes which are not divisors of ρ. The entire function

$$1 - \chi(p)p^{-s}$$

has its zeros on the imaginary axis. The function

$$p^{\frac{1}{2}s} - \chi(p)p^{-\frac{1}{2}s}$$

is a limit of polynomials whose zeros lie on the imaginary axis. Convergence is uniform on compact subsets of the complex plane. The Euler product converges uniformly on compact subsets of the half-plane $\mathbb{R}s > 1$. A construction of the space $\mathcal{H}(E)$ results from the Sonine spaces of order v for the complex plane. The main step in the construction will be indicated.

If a space $\mathcal{H}(E_0)$ has dimension greater than r and if $S(z)$ is a polynomial of degree r whose zeros lie at distance at least one from the upper half-plane, then a space $\mathcal{H}(E_r)$ exists, whose elements are the entire functions $F(z)$ such that $S(z)F(z)$ belongs to the space $\mathcal{H}(E_0)$, such that multiplication by $S(z)$ is an isometric transformation of the space $\mathcal{H}(E_r)$ into the space $\mathcal{H}(E_0)$. If a maximal dissipative transformation in the space $\mathcal{H}(E_0)$ is defined by taking $F(z)$ into $F(z+i)$ whenever the functions $F(z)$ and $F(z+i)$ of z belong to the space, then a maximal dissipative transformation in the space $\mathcal{H}(E_r)$ is defined by taking $F(z)$ into $G(z+i)$ whenever the functions $F(z)$ and $G(z+i)$ are elements of the space such that the identity

$$S(z)G(z + i) = H(z + i)$$

holds for an element $H(z)$ of the space $\mathcal{H}(E_0)$ with $P(z)F(z)$ nearest $H(z)$. Entire functions $P(z)$ and $Q(z)$ exist such that the function

$$\frac{Q(z)P(w^-) - P(z)Q(w^-)}{\pi(z - w^-)}$$

of z belongs to the space $\mathcal{H}(E_r)$ for every complex number w and such that the identity

$$G(w) = \left\langle F(t), \frac{Q(t)P(w^-) - P(t)Q(w^-)}{\pi(z - w^-)} \right\rangle_{\mathcal{H}(E_r)}$$

holds for all complex numbers w whenever the transformation in the space $\mathcal{H}(E_r)$ takes $F(z)$ into $G(z+i)$.

FURTHER READING

For anyone whose appetite has been whetted by this book, the following titles cover a number of aspects of pure mathematics in more detail. They all contain material that is appreciable by the non-mathematician.

Journey Through Genius: The Great Theorems of Mathematics,
William Dunham, John Wiley and Sons, 1990
Although this book doesn't deal with the Riemann Hypothesis, it gives a very good survey of how mathematicians have set about proving other important theorems.

A Mathematician's Apology, G. H. Hardy, Cambridge University Press, 1967
The gold standard of writing about mathematics for non-mathematicians.

Mathematical Mysteries, Calvin C. Clawson, Plenum Press, New York
and London, 1996
This book contains the best detailed explanation of the Riemann Hypothesis for the non-mathematician, as well as providing an excellent account of number theory in general.

The Man Who Knew Infinity, Robert Kanigel, Scribners, New York, 1991
An account of the life and work of Ramanujan.

The Development of Mathematics, E.T. Bell, Dover Publications Inc,
New York, 1992
Men of Mathematics, E. T. Bell, Simon and Schuster, New York, 1986
These two books are classics in the field. Although the first involves a lot of mathematical material it gives an insight into the philosophical underpinnings of mathematics which can be appreciated by anyone. The second contains fascinating biographies of many major mathematicians.

Littlewood's Miscellany, Edited by Bela Bollobas, Cambridge University Press,
1986
An eccentric and delightful selection of mathematical writings by Jack Littlewood. The introduction by Bollobas, a friend of Littlewood's, is excellent and most of the contents require no mathematical knowledge to enjoy.

Mathematical Expeditions, Reinhard Laubenbacher and David Pengelly,
Springer, New York, 1999
An undergraduate text that nevertheless contains many good biographical and historical insights into how mathematicians think and work.

INDEX